Voice of the Universe

Building the Jodrell Bank Telescope

REVISED AND UPDATED EDITION

BERNARD LOVELL

CONVERGENCE
A Series Founded, Planned, and Edited
by Ruth Nanda Anshen

PRAEGER

New York
Westport, Connecticut
London

Library of Congress Cataloging-in-Publication Data

Lovell, Bernard, Sir, 1913-
Voice of the universe.

(Convergence)
Rev. ed. of: The story of Jodrell Bank. 1968.
Includes index.
1. Nuffield Radio Astronomy Laboratories, Jodrell Bank—History. I. Title. II. Lovell, Bernard, Sir, 1913- . Story of Jodrell Bank. III. Series: Convergence (New York, N.Y.)
QB479.G72N86 1987 522'.19427'16 87-9322
ISBN 0-275-92678-8 (alk. paper)
ISBN 0-275-92679-6 (pbk. : alk. paper)

Library of Congress Catalog Card Number: 87-9322
ISBN: 0-275-92678-8
ISBN: 0-275-92679-6 (paperback)

First published in 1987

Praeger Publishers, 1 Madison Avenue, New York, NY 10010
A division of Greenwood Press, Inc.

Printed in the United States of America

The paper used in this book complies with the Permanent Paper Standard issued by the National Information Standards Organization (Z39.48-1984).

10 9 8 7 6 5 4 3 2 1

CONVERGENCE

A Series Founded, Planned, and Edited
by Ruth Nanda Anshen

BOOKS IN THE CONVERGENCE SERIES

The Double-Edged Helix
Genetic Engineering in the Real World
Liebe F. Cavalieri

Knowledge of Language
Its Nature, Origin, and Use
Noam Chomsky

Progress or Catastrophe
The Nature of Biological Science and Its
Impact on Human Society
Bentley Glass

Emerging Cosmology
Bernard Lovell

Creation and Evolution
Myth or Reality?
Norman D. Newell

Anatomy of Reality
Merging of Intuition and Reason
Jonas Salk

Science and Moral Priority
Merging Mind, Brain, and Human Values
Roger Sperry

Voice of the Universe
Building the Jodrell Bank Telescope
Bernard Lovell

CONTENTS

ILLUSTRATIONS

Illustrations are found following page 122.

IN TEXT

CONVERGENCE

RUTH NANDA ANSHEN

"THERE is no use trying," said Alice; "one *can't* believe impossible things."

"I dare say you haven't had much practice," said the Queen, "When I was your age, I always did it for half an hour a day. Why, sometimes I've believed as many as six impossible things before breakfast."

This commitment is an inherent part of human nature and an aspect of our creativity. Each advance of science brings increased comprehension and appreciation of the nature, meaning, and wonder of the creative forces that move the cosmos and created man. Such openness and confidence lead to faith in the reality of possibility and eventually to the following truth: "The mystery of the universe is its comprehensibility."

When Einstein uttered that challenging statement, he could have been speaking about our relationship with the universe. The old division of the Earth and the Cosmos into objective processes in space and time and mind in which they are mirrored is no longer a suitable starting point for understanding the universe, science, or ourselves. Science now begins to focus on the convergence of man and nature, on the framework which makes us, as living beings, dependent parts of nature and simultaneously makes nature the object of our thoughts and actions. Scientists can no longer confront the universe as objective observers. Science recognizes the participation of man with the universe. Speaking quantitatively, the universe is largely indifferent to what happens in man. Speaking qualitatively, nothing happens in man that does not have a bearing on the elements which constitute the universe. This gives cosmic significance to the person.

Nevertheless, all facts are not born free and equal. There exists a hierarchy of facts in relation to a hierarchy of values. To arrange the facts rightly, to differentiate the important from the trivial, to see their bearing in relation to each other and to evaluational criteria, requires a judgement which is intuitive as well as empirical. We need meaning in addition to information. Accuracy is not the same as truth.

Our hope is to overcome the cultural *hubris* in which we have been living. The scientific method, the technique of analyzing, explaining, and classifying, has demonstrated its inherent limitations. They arise because, by its intervention, science presumes to alter and fashion the object of its investigation. In reality, method and object can no longer be separated. The outworn Cartesian, scientific world view has ceased to be scientific in the most profound sense of the word, for a common bond links us all—man, animal, plant, and galaxy—in the unitary principle of all reality. For the self without the universe is empty.

This universe of which we human beings are particles may be defined as a living, dynamic process of unfolding. It is a breathing universe, its respiration being only one of the many rhythms of its life. It is evolution itself. Although what we observe may seem to be a community of separate, independent units, in actuality these units are made up of subunits, each with a life of its own, and the subunits constitute smaller living entities. At no level in the hierarchy of nature is independence a reality. For that which lives and constitutes matter, whether organic or inorganic, is dependent on discrete entities that, gathered together, form aggregates of new units which interact in support of one another and become an unfolding event, in constant motion, with ever-increasing complexity and intricacy of their organization.

Are there goals in evolution? Or are there only discernible patterns? Certainly there is a law of evolution by which we can explain the emergence of forms capable of activities which are indeed novel. Examples may be said to be the origin of life, the emergence of individual consciousness, and the appearance of language.

The hope of the concerned authors in Convergence is that they will show that evolution and development are interchangeable and that the entire system of the interweaving of man, nature, and the universe constitutes a living totality. Man is searching for his legitimate place in this unity, this cosmic scheme of things. The meaning of this cosmic scheme—if indeed we can impose meaning on the mystery and majesty of nature—and the extent to which we can assume responsibility in it as uniquely intelligent beings, are supreme questions for which this Series seeks an answer.

Inevitably, toward the end of a historical period, when thought and custom have petrified into rigidity and when the elaborate machinery of civilization opposes and represses our more noble qualities, life stirs again beneath the hard surface. Nevertheless, this attempt to define the purpose of Convergence is set forth with profound trepidation. We are living in a period of extreme darkness. There is moral atrophy, destructive radiation within us, as we watch the collapse of values hitherto cherished—but now betrayed. We seem to be face to face with an apocalyptic destiny. The anomie, the chaos, surrounding us produces an almost lethal disintegration of the person, as well as ecological and demographic disaster. Our situation is desperate. And there is no glossing over the deep and unresolved tragedy that fills our lives. Science now begins to question its premises and tells us not only what *is*, but what *ought* to be; *pre*scribing in addition to *de*scribing the realities of life, reconciling order and hierarchy.

My introduction to Convergence is not to be construed as a prefatory essay to each individual volume. These few pages attempt to set forth the general aim and purpose of this Series. It is my hope that this statement will provide the reader with a new orientation in his thinking, one more specifically defined by these scholars who have been invited to participate in this intellectual, spiritual, and moral endeavor so desperately needed in our time. These scholars recognize the relevance of the nondiscursive experience of life which the discursive, analytical method alone is unable to convey.

The authors invited to Convergence Series acknowledge a structural kinship between subject and object, between living and nonliving matter, the immanence of the past energizing the present and thus bestowing a promise for the future. This kinship has long been sensed and experienced by mystics. Saint Francis of Assisi described with extraordinary beauty the truth that the more we know about nature, its unity with all life, the more we realize that we are one family, summoned to acknowledge the intimacy of our familial ties with the universe. At one time we were so anthropomorphic as to exclude as inferior such other aspects of our relatives as animals, plants, galaxies, or other species—even inorganic matter. This only exposed our provincialism. Then we believed there were borders beyond which we could not, must not, trepass. These frontiers have never existed. Now we are beginning to recognize, even take pride in, our neighbors in the Cosmos.

Human thought has been formed through centuries of man's consciousness, by perceptions and meanings that relate us to nature. The smallest living entity, be it a molecule or a particle, is at the same time present in the structure of the Earth and all its inhabitants, whether human or manifesting themselves in the multiplicity of other forms of life.

Today we are beginning to open ourselves to this evolved experience of consciousness. We keenly realize that man has intervened in the evolutionary process. The future is continent, not completely prescribed, except for the immediate necessity to evaluate in order to live a life of integrity. The specific gravity of the burden of change has moved from genetic to cultural evolution. Genetic evolution itself has taken millions of years; cultural evolution is a child of no more than twenty or thirty thousand years. What will be the future of our evolutionary course? Will it be cyclical in the classical sense? Will it be linear in the modern sense? Yet we know that the laws of nature are not linear. Certainly, life is more than mere endless repetition. We must restore the importance of each moment, each deed. This is impossible if the future is nothing but a mechanical extrapolation of the past. Dignity becomes possible only with choice. The choice is ours.

In this light, evolution shows man arisen by a creative power inherent in the universe. The immense ancenstral effort that has borne man invests him with a cosmic responsibility. Michelangelo's image of Adam created at God's command becomes a more intelligent symbol of man's position in the world than does a description of man as a chance aggregate of atoms or cells. Each successive stage of emergence is more comprehensive, more meaningful, more fulfilling, and more converging, than the last. Yet a higher faculty must always operate through the levels that are below it. The higher faculty must enlist the laws controlling the lower levels in the service of higher principles, and the lower level which enables the higher one to operate through it will always limit the scope of these operations, even menacing them with possible failure. All our higher endeavors must work through our lower forms and are necessarily exposed thereby to corruption. We may thus recognize the cosmic roots of tragedy and our fallible human condition. Language itself as the power of universals, is the basic expression of man's ability to transcend his environment and to transmute tragedy into a moral and spiritual triumph.

This relationship, this convergence, of the higher with the lower applies again when an upper level, such as consciousness or freedom, endeavors to reach beyond itself. If no higher level can be accounted for by the operation of a lower level, then no effort of ours can be truly creative in the sense of establishing a higher principle not intrinsic to our initial condition. And establishing such a principle is what all great art, great thought, and great action must aim at. This is indeed how these efforts have built up the heritage in which our lives continue to grow.

Has man's intelligence broken through the limits of his own powers? Yes and no. Inventive efforts can never fully account for their success, but the story of man's evolution testifies to a creative power that goes beyond that which we can account for in ourselves. This power can make us surpass ourselves. We exercise some of it in the simple act of acquiring knowledge and holding it to be true. For, in doing so, we strive for intellectual control over things outside ourselves, in spite of our manifest incapacity to justify this hope. The greatest efforts of the human mind amount to no more than this. All such acts impose an obligation to strive for the ostensibly impossible, representing man's search for the fulfillment of those ideals which, for the moment, seem to be beyond his reach. For the good of a moral act is inherent in the act itself and has the power to ennoble the person who performs it. Without this moral ingredient there is corruption.

The origins of one person can be envisaged by tracing that person's family tree all the way back to the primeval specks of protoplasm in which his first origins lie. The history of the family tree converges with everything that has contributed to the making of a human being. This segment of evolution is on a par with the history of a fertilized egg developing into a mature person, or the history of a plant growing from a seed; it includes everything that caused that person, or that plant, or that animal, or even that star in a galaxy, to come into existence. Natural selection plays no part in the evolution of a single human being. We do not include in the mechanism of growth the possible adversities which did not befall it and hence did not prevent it. The same principle of development holds for the evolution of a single human being; nothing is gained in understanding this evolution by considering the adverse chances which might have prevented it.

In our search for a reasonable cosmic view, we turn in the first place to common understanding. Science largely relies for its subject matter on a common knowledge of things. Concepts of life and death, plant and animal, health and sickness, youth and age, mind and body, machine and technical processes, and other innumerable and equally important things are commonly known. All these concepts apply to complex entities, whose reality is called into question by a theory of knowledge which claims that the entire universe should ultimately be represented in all its aspects by the physical laws governing the inanimate substrate of nature. "Technological inevitability" has alienated our relationship with nature, with work, with other human beings, with ourselves. Judgment, decision, and freedom of choice, in other words *knowledge* which contains a moral imperative, cannot be ordered in the form that some technological scientists believe. For there is no mechanical ordering, no exhaustive set of permutations or combinations that can perform the task. The power which man has achieved through technology has been transformed into spiritual and moral impotence. Without the insight into the nature of *being*, more important than *doing*, the soul of man is imperilled. And those self-transcendent ends that ultimately confer dignity, meaning and identity on man and his life constitute the only final values worth pursuing. The pollution of consciousness is the result of mere technological efficiency. In addition, the authors in this Series recognize that the computer in itself can process information—not meaning. Performance is now substituted for thought; the doer substituted for the thinker. Thus we see on the stage of life no moral actors, only anonymous events.

Our new theory of knowledge, as the authors in this Series try to demonstrate, rejects this claim and restores our respect for the immense range of common knowledge acquired by our experience of convergence. Starting from here, we sketch out our cosmic perspective by exploring the wider implications of the fact that all knowledge is acquired and possessed by relationship, coalescense, convergence.

We identify a person's physiognomy by depending on our awareness of features that we are unable to specify, and this amounts to a convergence in the features of a person for the purpose of comprehending their joint meaning. We are also able to read in the features and behavior of a person the presence of moods, the gleam of intelligence, the response to animals or a sunset or a fugue by Bach, the signs of sanity, human responsibility, and experience. At a lower level, we comprehend by a similar mechanism the body of a person and understand the functions of the physiological mechanism. We know that even physical theories constitute in this way the processes of inanimate nature. Such are the various levels of knowledge acquired and possessed by the experience of convergence.

The authors in this Series grasp the truth that these levels form a hierarchy of comprehensive entities. Inorganic matter is comprehended by physical laws; the mechanism of physiology is built on these laws and enlists them in its service. Then, the intelligent behavior of a person relies on the healthy functions of the body and, finally, moral responsibility relies on the faculties of intelligence directing moral acts.

We realize how the operations of machines, and of mechanisms in general, rely on the laws of physics but cannot be explained, or accounted for, by these laws. In a hierarchic sequence of comprehensive levels, each higher level is related to the levels below it in the same way as the operations of a machine are related to the particulars, obeying the laws of physics. We cannot explain the operations of an upper level in terms of the particulars on which its operations rely. Each higher level of integration represents, in this scene, a higher level of existence, not completely accountable by the levels below it yet including these lower levels implicitly.

In a hierarchic sequence of comprehensive levels each higher level is known to us by relying on our awareness of the particulars on the level below it. We are conscious of each level by internalizing its particulars and mentally performing the integration that constitutes it. This is how all experience, as well as all knowledge, is based on convergence, and this is how the consecutive stages of convergence form a continuous transition from the understanding of the inorganic, the inanimate, to the comprehension of man's moral responsibility and participation in the totality, the organismic whole, of all reality. The sciences of the subject-object relationship thus pass imperceptibly into the metascience of the convergence of the subject and object interrelationship, mutually altering each other. From the minimum of convergence, exercised in a physical observation, we move without a break to the maximum of convergence, which is a total commitment.

"The last of life, for which the first was made, is yet to come." Thus, Convergence has summoned the world's most concerned thinkers to rediscover the experience of *feeling*, as well as of thought. The convergence of all forms of reality presides over the possible fulfillment of self-awareness—not the isolated, alienated self, but rather the participation in the life process with other lives and other forms of life. Convergence is a cosmic force and may possess liberating powers allowing man to become what he is, capable of freedom, justice, love. Thus man experiences the meaning of grace.

A further aim of this Series is not, nor could it be, to disparage science. The authors themselves are adequate witness to this fact. Actually, in viewing the role of science, one arrives at a much more modest judgment of its function in our whole body of knowledge. Original knowledge was probably not acquired by us in the active sense; most of it must have been given to us in the same mysterious way we received our consciousness. As to content and usefulness, scientific knowledge is an infinitesimal fraction of natural knowledge. Nevertheless, it is knowledge whose structure is endowed with beauty because its abstractions satisfy our urge for specific knowledge much more fully than does natural knowledge, and we are justly proud of scientific knowledge because we can call it our own creation. It teaches us clear thinking, and the extent to which clear thinking helps us to order our sensations is a marvel which fills the mind with ever new and increasing admiration and awe. Science now begins to include the realm of human values, lest even the memory of what it means to be human be forgotten. In fact, it may well be that science has reached the limits of the knowable and may now be required to recognize its inability to penetrate into the caprice and the mystery of the soul of the atom.

Organization and energy are always with us, wherever we look, on all levels. At the level of the atom organization becomes indistinguishable from form, from order, from whatever the forces are that held the spinning groups of ultimate particles together in their apparent solidity. And now that we are at the atomic level, we find that modern physics has recognized that these ultimate particles are primarily electrical charges, and that mass is therefore a manifestation of energy. This has often been misinterpreted by idealists as meaning that matter has somehow been magicked away as if by a conjuror's wand. But nothing could be more untrue. It is impossible to transform matter into spirit just by making it thin. Bishop Berkeley's views admit of no refutation but carry no conviction nevertheless. However, something has happened to matter. It was only separated from form because it seemed too simple. Now we realize that, and this is a revolutionary change; we cannot separate them. We are now summoned to cease speaking of Form and Matter and begin to consider the convergence of Organization and Energy. For the largest molecule we know and the smallest living particles we know overlap. Such a cooperation, even though far down at the molecular level, cannot but remind us of the voluntary cooperation of individual human beings in maintaining patterns of society at levels of organization far higher. The tasks of Energy and Organization in the making of the universe and ourselves are far from ended.

No individual destiny can be separated from the destiny of the universe. Alfred North Whitehead has stated that every event, every step or process in the universe, involves both effects from past situations and the anticipation of future potentialities. Basic for this doctrine is the assumption that the course of the universe results from a multiple and never-ending complex of steps developing out of one another. Thus, in spite of all evidence to the contrary, we conclude that there is a continuing and permanent energy of that which is not only man but all life. For not an atom stirs in matter, organic and inorganic, that does not have its cunning duplicate in mind. And faith in the convergence of life with all its multiple manifestations creates its own verification.

We are concerned in this Series with the unitary structure of all nature. At the beginning, as we see in Hesiod's *Theogony* and in the Book of Genesis, there was a primal unity, a state of fusion in which, later, all elements become separated but then merge again. However, out of this unity there emerge, through separation, parts of opposite elements. These opposites intersect or reunite, in meteoric phenomena or in individual living things. Yet, in spite of the immense diversity of creation, a profound underlying convergence exists in all nature. And the principle of the conservation of energy simply signifies that there is a *something* that remains constant. Whatever fresh notions of the world may be given us by future experiments, we are certain beforehand that something remains unchanged which we may call *energy*. We now do not say that the law of nature springs from the invariability of God, but with that curious mixture of arrogance and humility which scientists have learned to put in place of theological terminology, we say instead that the law of conservation is the physical expression of the elements by which nature makes itself understood by us.

The universe is our home. There is no other universe than the universe of all life including the mind of man, the merging of life with life. Our consciousness is evolving, the primordial principle of the unfolding of that which is implied or contained in all matter and spirit. We ask: Will the central mystery of the cosmos, as well as man's awareness of and participation in it, be unveiled, although forever receding, asymptotically? Shall we perhaps be able to see all things, great and small, glittering with new light and reborn meaning, ancient but now again relevant in an iconic image which is related to our own time and experience?

The cosmic significance of this panorama is revealed when we consider it as the stages of an evolution that has achieved the rise of man and his consciousness. This is the new plateau on which we now stand. It may seem obvious that the succession of changes, sustained through a thousand million years, which have transformed microscopic specks of protoplasm into the human race, has brought forth, in so doing, a higher and altogether novel kind of being capable of compassion, wonder, beauty and truth, although each form is as precious, as sacred, as the other. The interdependence of everything with everything else in the totality of being includes a participation of nature in history and demands a participation of the universe.

The future brings us nothing, gives us nothing; it is we who in order to build it have to give it everything, our very life. But to be able to give, one has to possess; and we possess no other life, no living sap, than the treasures stored up from the past and digested, assimilated, and created afresh by us. Like all human activities, the law of growth, of evolution, of convergence draws its vigor from a tradition which does not die.

At this point, however, we must remember that the law of growth, of evolution, has both a creative and a tragic nature. This we recognize as a degenerative process, as devolution. Whether it is the growth of a human soul or the growth of a living cell or of the universe, we are confronted not only with fulfillment but with sacrifice, with increase and decrease, with enrichment and diminution. Choice and decision are necessary for growth and each choice, each decision, excludes certain potentialities, certain potential realities. But since these unactualized realities are part of us, they possess a right and command of their own. They must avenge themselves for their exclusion from existence. They may perish and with them all the potential powers of their existence, their creativity. Or they may not perish but remain unquickened within us, repressed, lurking, ominous, swift to invade in some disguised form our life process, not as a dynamic, creative, converging power, but as a necrotic, pathological force. If the diminishing and the predatory processes comingle, atrophy and even death in every category of life ensue. But if we possess the maturity and the wisdom to accept the necessity of choice, of decision, or order and hierarchy, the inalienable right of freedom and autonomy, then, in spite of its tragedy, its exclusiveness, the law of growth endows us with greatness and a new moral dimension.

Convergence is committed to the search for the deeper meanings of science, philosophy, law, morality, history, technology, in fact all the disciplines in a trans-disciplinary frame of reference. This Series aims to expose the error in that form of science which creates an unreconcilable dichotomy between the observer and the participant, thereby destroying the uniqueness of each discipline by neutralizing it. For in the end we would know everything but *understand nothing*, not being motivated by concern for any question. This Series further aims to examine relentlessly the ultimate premises on which work in the respective fields of knowledge rests and to break through from these into the universal principles which are the very basis of all specialist information. More concretely, there are issues which wait to be examined in relation to, for example, the philosophical and moral meanings of the models of modern physics, the question of the purely physico-chemical processes versus the postulate of the irreducibility of life in biology. For there is a basic correlation of elements in nature, of which man is a part, which cannot be separated, which compose each other, which converge, and alter each other mutually.

Certain mysteries are now known to us: the mystery, in part, of the universe and the mystery of the mind have been in a sense revealed out of the heart of darkness. Mind and matter, mind and brain, have converged; space, time, and motion are reconciled; man, consciousness, and the universe are reunited since the atom in a star is the same as the atom in man. We are homeward bound because we have accepted our convergence with the Cosmos. We have reconciled observer and participant. For at last we know that time and space are modes by which we think, but not conditions in which we live and have our being. Religion and science meld; reason and feeling merge in mutual respect for each other, nourishing each other, deepening, quickening, and enriching our experiences of the life process. We have heeded the haunting voice in the Whirlwind.

THE MÖBIUS STRIP

THE symbol found on the cover of each volume in Convergence is the visual image of *convergence*—the subject of this Series. It is a mathematical mystery deriving it names from Augustus Möbius, a German mathematician who lived from 1790 to 1868. The topological problem still remains unsolved mathematically.

The Möbius Strip has only one continuous surface, in contrast to a cylindrical strip, which has two surfaces—the inside and the outside. An examination will reveal that the Strip, having one continuous edge, produces *one* ring, twice the circumference of the original Strip with one half of a twist in it, which eventually *converges with itself*.

Since the middle of the last century, mathematicians have increasingly refused to accept a "solution" to a mathematical problem as "obviously true," for the "solution" often then becomes the problem. For example, it is certainly obvious that every piece of paper has two sides in the sense that an insect crawling on one side could not reach the other side without passing around an edge or boring a hole through the paper. Obvious—but false!

The Möbius Strip, in fact, presents only one mono-dimensional, continuous ring having no inside, no outside, no beginning, no end. Converging with itself it symbolizes the structural kinship, the intimate relationship between subject and object, matter and energy, demonstrating the error of any attempt to bifurcate the observer and participant, the universe and man, into two or more systems of reality. All, all is unity.

I am indebted to Fay Zetlin, Artist-in-Resident at Old Dominion University in Virginia, who sensed the principle of convergence, of emergent transcendence, in the analogue of the Möbius Strip. This symbol may be said to crystallize my own continuing and expanding explorations into the unitary structure of all reality. Fay Zetlin's drawing of the Möbius Strip constitutes the visual image of this effort to emphasize the experience of convergence.

R.N.A.

INTRODUCTION TO THE REVISED AND UPDATED EDITION

THE first thirty-seven chapters of this book were originally published in 1968 as *The Story of Jodrell Bank*. That book told the story of the development of a new technique for studying the universe—radio astronomy—from the moment in 1945 when I arrived with two ex-army trailers of radar equipment at a field in Cheshire, England, twenty-five miles south of the city of Manchester.

I wanted to build a new kind of telescope—a radio telescope—of great size which could be steered to receive signals from any part of the heavens. At last I was successful, but it was a success very close indeed to disaster. During the five years of construction I had become entwined in a formidable series of engineering, political and financial problems to such an extent that in the days of completion I felt that only a miracle could save me and the Jodrell establishment from extinction. Then, during an unforgettable night in October 1957, that miracle came when we located by radar the carrier rocket of the first Soviet Sputnik. It was a rocket, which in another guise, could launch a missile. The realisation that we had built a telescope with such a capability transformed overnight public and private attitudes to my problems. Three years later an American finger pressed a switch in a trailer near the telescope that transmitted a signal from the telescope to release the American Pioneer V space probe from its carrier rocket a few minutes after it had been launched from Cape Canaveral. That event, which inspired a gift of such magnitude that our outstanding debt was cleared, was the end of the original story.

Indeed, it was ironical that our salvation came from a use of the telescope for which it was not primarily intended. It was built to study the universe with a new technique and that has been the main use of the telescope for nearly thirty years. When I wrote the original Introduction in 1967 the telescope was ten years old. I had argued that it would be a useful instrument for fifteen years. Even in my moments of greatest optimism I do not think that in those days I would have believed that after thirty years it would still be working as a major astronomical instrument in the international scene. *The Story of Jodrell Bank* has for long been out of print and it is a very great pleasure for me to learn that it is now to be re-issued. However, the end of the original book would inevitably leave the reader wondering what had happened to Jodrell Bank and the telescope during the subsequent quarter of a century. I have therefore added three more chapters to answer that query and to describe a little of the astronomical researches into the nature of the universe.

The suggestion that *The Story of Jodrell Bank* should be re-published with this additional material under the title *Voice of the Universe* was made by Dr. Ruth Nanda Anshen during her visit to Jodrell Bank in the summer of 1986. I am very deeply indebted to Dr. Anshen for making this possible by including the book in the Convergence series which she has founded, planned and edited with such insight.

Note

Inflation has considerably changed the purchasing value of the pound sterling during the years covered in this book. To convert the figures given in this book to 1986 values an approximate guide may be estimated as follows:

1950	multiply	by	11
1955	"	"	9
1960	"	"	8
1965	"	"	7
1970	"	"	5.5
1975	"	"	3
1980	"	"	1.5

The equivalent U.S. dollar exchange rate was approximately £1 sterling = 2.8 dollars until the devaluation of 1967. During the 1969-71 period of Chapter 39 the exchange rate was approximately £1 sterling = 2.4 dollars. During the period covered in Chapter 40 the rate varied from £1 = 1.66 dollars (1976) to £1 = 2.38 dollars (1980).

Bernard Lovell
Jodrell Bank
1986

FOREWORDS

by

H. C. Husband, *C.B.E., D.Sc., Consulting Engineer*

and

R. A. Rainford, *M.A., F.A.C.C.A., Bursar, University of Manchester*

An author must be allowed to write a book in his own way, but it is unusual for a character to be taken from the plot and instructed to write a foreword.

This is the story of acquiring and spending a large sum of money on equipment for the advancement of pure science. It has always been difficult for most people to understand the need for scientific research unless there is an immediate object in view. Only the devoted few were originally in favour of Professor Lovell's great telescope, and perhaps most of these had no very clear idea as to how it was to be used. Indeed, many new applications have been revealed during the past ten years.

There might have been no happy ending to the story, financially at any rate, had not radio astronomy become entangled, not at all to the author's liking, with the mechanics of space research, an expression a little more self-explanatory. The telescope immediately became popular and eventually a symbol of Britain's efforts in the 'space race' because of the fortunate timing of the launch of the first Sputnik, just when the radio astronomers were in the depths of despair. Jodrell Bank then, to its horror, almost overnight found itself dealing in applied science, which is something more tangible to those responsible for spending the country's money.

The history of the Jodrell Bank troubles has been compiled very largely from the diary kept by Professor Lovell, 'composed from day to day under the stress of events and with the knowledge available at the moment'. Life is thereby given to the story. I have a strong feeling that some of the opinions and criticisms copied from the daybook are different from those which might have been made in retrospect—it must be very difficult to be a historian of one's own affairs—but the story seems to me better for describing both the controversies and the emotions of the moment.

No work with which I have ever been associated was carried out by more devoted contractors, yet so eagerly was the completion of the telescope desired that any delay during the construction provoked rage rather than excited sympathy; but the fascination of Jodrell Bank has prevailed and those who contributed to the work there in the early days have returned more than once to assist in later projects.

I feel there is an acknowledgement which the author has omitted and which he may allow me to mention. We were both fortunate during those years of considerable stress and anxiety to be troubled with no domestic worries. Lady Lovell has a remarkable calmness and sense of humour, and my wife was, to say the least, tolerant of my almost constant absence from home. Both continued the work of bringing up our large families whilst we went about the building of the first really big radio telescope.

H.C.H.

It was with pleasure that I heard that Sir Bernard Lovell had written an account of the birth and development of the Manchester University Radio Astronomy Department at Jodrell Bank and the problems surrounding its early days.

I have read the manuscript and, inevitably, have lived again those anxious and exciting times through which we passed before the largest steerable radio telescope (Mark I as it is affectionately called) was fully operational and finally paid for.

The project could never have been carried to its successful conclusion without the fanatical enthusiasm, drive, and energy of Bernard Lovell, and only those near to him appreciate the strain which he underwent during those days.

Enthusiasm is infectious and when it is allied to outstanding ability the result is certain. It is impossible to work with Bernard Lovell without being affected and I, like most who knew him well, was carried along by this enthusiasm. Even in the darkest hours when quick decisions had to be made—decisions which one knew would lead to later inquests—it was the confidence in his ability to make a success of the radio telescope when completed that made the University press on with all the speed it could. Events have proved the University right and it is now praised for the very acts for which it was once criticized.

It is extremely doubtful if such a venture could ever have been completed so effectively except by a University. Universities back men and ideas and it is only in such a community where one can know men well and where ideas may be quickly assessed that such new sciences as radio astronomy can come to the fore rapidly. Government departments are good at supporting established disciplines and matters of 'timeliness and promise' but they must, by their very nature, be ultra-cautious when dealing impersonally at long range with new ventures. This is not a criticism of Government departments as I do not see how they could act otherwise, but it is an argument for local initiative and central encouragement.

I have always seen the birth of the Radio Astronomy Department of Manchester University as an example of the value of University freedom. There are, of course, many other examples but I hope that the readers of this book will consider the account of this enterprise a proof of the value of giving men with original ideas an opportunity of pursuing their work with the backing of their colleagues. The opening remarks of Chapter 2 ('Growth') will give the reader an impression of what I mean.

Even though it is the working of the Mark I radio telescope which has gripped the imagination of ordinary folk, I, personally, will never forget the first large aerial system which was built by Lovell and Clegg and a few assistants in 1947, and which is described in Chapter 4. The results of their efforts, intellectual and physical, can hardly be exaggerated and it was these results which led to the idea of a large movable 'dish' and the association with Dr. Husband, who had the courage and engineering skill to tackle a job which would have deterred most engineers.

It was ironical that when the radio telescope was proving its worth, but the University was still being criticized, the then Prime Minister (the Rt. Hon. Harold Macmillan) in the House of Commons said: 'Hon. members will have seen that within the last few days *our great radio telescope at Jodrell Bank* has successfully tracked the Sputnik's carrier rocket'. The italics are mine and when I read this report in the press I felt he ought to have added after 'Jodrell Bank', the words: 'which was completed in spite of us'.

It has always been an annoyance to me that whenever the success of the radio telescope is mentioned it is referred to as the Jodrell Bank Research Station as though it were a part of a Government or independent establishment, but whenever it was criticized in the early days the name of the University of Manchester was invariably used.

Some time ago Sir Bernard asked me if, with my present knowledge, I would go through those trying times again and I answered that I did not know, though I knew I would, as my faith in him and his colleagues never faltered and events proved them so right.

I am proud to have been associated with Lovell and Husband in this great project and to know that work has been done and is being done on this instrument which redounds to the credit of the country.

I must correct one, for me, fatal omission in the book. Lovell tells of the work of many but there is little reference to one person whose activity during the most nerve-racking and frustrating period was continuous. I refer to Lady Lovell whose invaluable but unobtrusive help was not only a source of strength to Bernard Lovell but to all of us. Her common sense, loyalty, good humour, ready wit, and supreme confidence were incalculable. It is in keeping with her character that her husband is not allowed to mention her but although I am risking her friendship, I cannot let this opportunity pass without paying my tribute to her and offering sincere thanks.

R.A.R.

INTRODUCTION

THIS is the story of the emergence of Jodrell Bank as a scientific establishment from a foggy day in December 1945 when my trailer of ex-Army radar equipment was towed into a muddy field in Cheshire. As I thawed myself before a coke stove in a gardener's hut on that day I had no vision of the future other than to restart my scientific career after the frustrating years of the war. If I could have seen a flash to an October day twelve years later with the world's pressmen packed in a room under a huge radio telescope then I would have dismissed it as a dream fantasy. Or if the crackle of the coke had changed to the chatter of two adjacent teleprinters, one connected to Moscow and the other to Washington, then I could only have thought that these were idle meanderings of a scientist's brain. An amazing and unpredictable conjunction of historic events was to bring such fantasies to life. Less than fourteen years after that December day the telescope recorded the impact of a Russian rocket on the moon. The story ends fifteen years later, in 1960, when, after a signal from the telescope had released the American space probe Pioneer V from its carrier rocket, a telephone call from Lord Nuffield conveyed the joyous news that we were at last extracted from the frightful entanglements which we had unwittingly created for ourselves. In that springtime the University freed itself of the huge debt for which I had been mainly responsible, and which had almost destroyed those of us associated with the building of the great radio telescope. It is unlikely that a scientific project has ever survived so many crises to attain ultimate success.

I realize that this story of the telescope is merely my own story. Others, whose names occur so frequently in these pages—particularly Dr. H. C. Husband and Mr. R. A. Rainford—could equally well describe those years as seen through different eyes. My story is that of a person for whom Jodrell Bank and the telescope became an obsession to be realized at any cost. Over those years it occupied almost the whole of my life and thought and hence, inevitably, the story is sometimes almost autobiographical. Furthermore, there was the peculiar circumstance through which I dictated an almost daily diary from the moment in 1952 when I knew we could move ahead to the time in 1957 when my career was in peril and Rainford remarked that my diary would either save me or finally destroy my career. I have never kept a diary before or since, but without the detail contained in it I could not have written this book.

I doubt, too, if the book would have been written if it had not been for the many friends and colleagues associated with those years who have urged me to do so. They felt that the extraordinary circumstances and complications surrounding the project demanded a coherent record. When we were free to move forward once more, other telescopes were built but the lessons of the first had been learnt and these smaller instruments were constructed without undue incident. They form no part of the present story.

I agreed to write this book in 1962, but for a person who is still young enough to work in the present and plan for the future it is hard to dwell in the past. I did almost nothing about if for two years and then an equinoctial gale created a strange circumstance which made me begin. In early October I drove to Southampton to cross the Atlantic in the s.s. *France*. Arriving in the early evening I found that the ship was still held in a French port by a severe gale and could not arrive until 3 a.m. at the earliest. Released from the possibility of either sleep or work I found a quiet corner of the terminal building and wrote Chapter 1 and I continued to write during the next few weeks in various parts of America. Indeed, I found even in that phantasmagoric creation of the modern world, Chicago airport on a Sunday afternoon, it is possible to find a place where the loudspeakers do not constantly search out individuals. In this strange fashion the first eight chapters came to life, then during Chapter 9 I had to put it away because I had reached the stage where I needed the dates and details in the diary. I wrote the remainder of the book in my office at Jodrell Bank where I had the diary, the files, and all the other documents which made it possible to establish facts which had been dimmed in the memory. A contemporary office, even of an astronomer, is a distracting room, and another three years passed and another two telescopes were built around me before I reached the end.

This book takes the story of Jodrell Bank to 1960, when it might be said that we became a reasonably normal scientific establishment. At that time, with the clearance of our debt, we became free to use the great telescope on the astronomical researches of our choice. It worked magnificently. The small group of young men who had stood by me during the years of crisis now applied their talents to the exploration of the universe. Soon, a stream of discoveries, vital to our understanding of the cosmos, began to flow from the recording instruments of the telescope.

In our age, science and its techniques have advanced with an almost alarming speed. Few of the contemporary large instruments of science remain effective on an international scale twenty years after their design. The Jodrell telescope has remained unique and unparalleled by the passage of time. No one has yet succeeded in building a larger one.

During the years of construction of the large telescope, space research and military requirements gave a great stimulus to the development of electronic techniques for use in the range of wavelengths shorter than those regarded as useful for radio astronomy when the Jodrell telescope was designed. Later telescopes have, on the whole, tended to sacrifice size for accuracy to enable them to study the universe in these centimetric wavelengths. It is a measure of the success and versatility of the Jodrell instrument that it has been able to absorb many of these new techniques. In spite of the enormous development of the subject it has remained the world's superior instrument in the wavelength ranges longer than about half a metre and has retained a competitive standing with more modern instruments at shorter wavelengths down to about 18 cm.

It was fortunate that, in spite of the troubles described in this book, the success of the telescope quickly predisposed the Department of Scientific and Industrial Research (and subsequently its successor, the Science Research Council) to listen with a sympathetic ear to my proposals for new developments at Jodrell Bank. In 1960 when this story ends I was already asking for money to build more radio telescopes. Two major ones have been built: one to enable us to explore the universe in the very short wavelength range, below the working limit of the large telescope, and a second to enable us to develop the technique of using another large telescope connected to the main Jodrell instrument by radio link for the measurement of the angular sizes of radio galaxies situated in the remote parts of the universe. These two instruments are now working. In their maximum dimension they are only half the size of the original telescope, but together and with their associated equipment they cost as much. The building of these instruments was completed without exceptional difficulty or associated trauma of the type described in this book. The more accurate of these smaller telescopes (known as the Mark II to distinguish it from the 250-ft. aperture (Mark I) telescope which is the subject of this book) was finished in 1964, and although it uses the same servo system of control as the 250-ft. telescope, a modern digital computer is used to feed instructions to this system. Instructions for telescope movement for a whole night are fed in on a yard of punched tape, and the results of the investigation appear also in punched tape form. It was the first instrument in the world to be driven 'on-line' by a digital computer in this manner. A similar digital system has recently been installed as an adjunct to the less accurate analogue computer of the Mark I.

In these ways Jodrell Bank has developed. The establishment and the large telescope have survived and absorbed the technical developments of twenty years. No one questions for a moment that in another twenty years the great telescope will still be working as one of the major scientific instruments in the world. It is unlikely to be the largest by then—for we ourselves are actively designing a larger one.

BERNARD LOVELL
Jodrell Bank

1

The origins of Jodrell Bank

SIX centuries ago a Cheshire archer, William Jauderell, distinguished himself fighting with the Black Prince at Poitiers. While he was thus engaged Jauderell learnt that his house in England had been destroyed by fire and he desired to return to attend to his affairs. The pass, inscribed on a roll of parchment from which hung a great seal, which the Black Prince gave him to make his journey possible, is still in the hands of the Jodrell family—for over the centuries the name Jauderell changed to Jodrell. Bank is the Cheshire name for a small rise or hill and thus the name Jodrell Bank is the description of a few acres of fields above a small brook lying twenty-five miles south of the city of Manchester near the village of Lower Withington. Nowadays thousands traverse it daily as they drive southwards from Manchester to join the M6 motorway near Holmes Chapel.

The region had little to distinguish it from the surrounding Cheshire plain—Jodrell Hall, which in modern times has become a preparatory school under a different name, a number of farms and smallholdings and at the foot of the bank the Red Lion Brook wending its way towards the River Dane near Holmes Chapel. It was here, six centuries after the return of William Jauderell from Poitiers and just before the outbreak of another great war, that the University of Manchester bought eleven acres of farmland from Ted Moston—one of the local farmers. The land was purchased for the horticultural botany department of the University.

At the time of this purchase in 1939 I was a young man teaching physics in the University in blissful ignorance of how my own life was to become entwined with Jodrell Bank, or with Moston and his tractor. During the war I remained unaware of Jodrell Bank although I must have driven past this horticultural botany department on many occasions when travelling from the south to visit aircraft firms in Manchester.

Of course, the war immediately wrenched almost every scientist in England from his research work, but with the incurable optimism of youth I continued to dream even on the freezing aerodromes of how I could continue my cosmic ray research when peace came. The dreams were stimulated by the strange echoes which so often appeared on the radar screens with which we were concerned—echoes which seemed to hold promise of a new technique for detecting large cosmic ray showers. These dreams would have remained ethereal but for one of the determinant things in my life—that I had been an Assistant Lecturer in a department where the head was P. M. S. Blackett,[1] the Langworthy Professor of Physics in Manchester, the successor of Schuster, Rutherford, and Bragg. Patrick Blackett's own wartime duties often led him to stay with us and the first result of the dreams was the Blackett–Lovell paper on the possibility of obtaining radar echoes from the ionization caused by large cosmic ray showers which was published in the Royal Society Proceedings during the war.[2]

Then the German armies surged across Europe in the spring of 1940 and all thoughts of cosmic ray research by old or new techniques became forgotten in the basic struggle for mere survival.

In the autumn of 1945 I returned to my academic career in Manchester, accompanied by two trailers of radar equipment and a portable diesel generator—equipment which only a few months earlier had been used to direct ack-ack guns to enemy aircraft. The appearance of this in the quadrangle outside the physics department caused consternation, but in the event it did not stay there for long. As soon as we got it working, hoping to begin our search for cosmic ray echoes, the cathode ray tube was covered with spikes from electrical interference. In those days electric trams were still running past the University, but even in the early hours of a Sunday morning when all was quiet the interference was far too serious for us to conceive of the possibility of research under these conditions.

I cannot remember who suggested that I should ask the University Bursar, R. A. Rainford, if the University had any land outside the city to which I could take my trailers. The next day I saw him in the Staff House, and that afternoon followed his directions to a field in the area of High Legh, near Lymm to the west of Manchester. It was hopeless—high-voltage grid lines traversed the land. I returned to the attack and little did he realize what troubles and near disasters his next remark was to lead us into!

[1] P. M. S. Blackett, O.M., C.H., F.R.S., Nobel Laureate; left Manchester for Imperial College, London, in 1953. He was elected President of the Royal Society in 1965.

[2] I sent a draft to Blackett which he tore up and rewrote in an air-raid shelter in Westminster during a London blitz. He wrote down the cross-section of scattering for an electron as $(e^2/mc^2)^2$ and it was a long time before he forgave me for failing to correct this by putting in an $\frac{8}{3}\pi$. It was useless for me to protest that I was very busy on urgent radar work, that I was 200 miles from a library and that it was a minor factor anyhow.

'You see that man with the beard over there drinking beer? That's Sansome,[1] the botanist, who runs a place called Jodrell Bank—try him.'

Sansome was full of enthusiasm—the exchange was perfect: although by profession a botanist he was also a tremendously keen radio experimenter; on the other hand I had the radio gear, but as an amateur I was already growing most of his plants in my garden. Indeed there were some occasions when I felt that we were better at each other's job.

In any case in early December my trailers were being towed by an Army convoy along the road south from Manchester to Jodrell Bank. The Army deposited me and the trailers and departed. I was at Jodrell Bank, with two wooden huts full of fertilizers and spades, two friendly gardeners—Alf Dean and Frank Foden—no electric power within miles, and in thick fog and hard frost.

I don't think that anything would have happened at Jodrell Bank but for Alf and Frank. The diesel had to be wound with a handle like a car starting-handle, but vastly heavier, and alone I could scarcely turn it. With the three of us together we could turn it at a speed at which the diesel should have sprung into life. Alas it remained cold and silent. But they knew Moston who had a tractor and a vastly inquiring mind. Soon the diesel was in pieces and the trouble was simply that it was so cold that the fuel pipes had become blocked with ice. On the third day the diesel started: electricity was being generated at Jodrell Bank, and I was able to switch on my radar transmitter and receiver. I still have the notebook with the entry for that day. It was 14 December, the date of maximum of the Geminid meteor shower. Then I knew nothing about meteors or Geminids and had no inkling that for many years subsequently 14 December and the Geminids were to be periods of intense activity at Jodrell Bank.

Alf and Frank were the sole occupants of Jodrell Bank and the two huts. In one of the huts there was a coke stove before which we thawed, brewed tea and ate our packed lunches. So, Jodrell Bank began. These are the answers to the frequent question 'Why did you pick on Jodrell Bank?' It picked me; the gardens, the people, the isolation were irresistible. For a short time, alas a very short time, I was happy and content with the trailers, Alf and Frank, and the coke stove.

[1] F. W. Sansome, C.B.E., subsequently Professor of Botany in the Ahmadu Bello University, Nigeria.

2

Growth

THE transformation of two trailers and a diesel generator in a field into a research laboratory happened for two reasons. Many men in the position of Patrick Blackett might have allowed a member of staff to engage in a short field excursion, but few would have encouraged the growth by the supply of money and assistance as he did. Even fewer would have done so when they realized that the primary object of the exercise had been abandoned in favour of other researches. But he was as excited as I was with the successive discoveries which soon poured into the trailers in the Cheshire fields and the idea of any retrenchment never occurred to him.

The other reason lies in the traditional freedom of University researches. No one except Blackett had any power to stop me doing whatever research I wanted with the apparatus in my possession. There was no formal proposal for research into the possibility of obtaining radar echoes from cosmic ray showers and no one to check that whatever time or money I spent would be on that problem. There was no commitment to write or give verbal reports, no fear that the researches would cease if the original idea failed. In fact, I never pursued the original plan of research; too many other exciting astronomical events impressed themselves on my screens. Many years later, when the original calculations were refined, it was clear that the experiment could not have succeeded—at least not with the apparatus at my disposal then.

The strange short-lived sporadic echoes of the war years were certainly present again on the cathode ray tube in the Cheshire field. Now there was time to absorb the literature of the past and opportunity to see a secret wartime report made by Dr. J. S. Hey of the Army Operational Research Group. There were strong indications that the transient echoes, most of which lasted for only a fraction of a second, were associated with meteors burning up in the earth's atmosphere. Hey and the A.O.R.G. had become associated with the problem in a strange manner. When the Germans started bombarding London with the V2 rockets the G.L. (gun laying) Radar Sets of A.A. Command were hastily pressed into service to detect the radar reflections from the rocket during its flight in order to give some minutes' warning of its arrival. These G.L. sets apparently carried out this detection task in an over-enthusiastic manner because many alarms were given when no rocket arrived. It was soon found that the operators were seeing echoes and giving the alarms when no rockets whatsoever were in process of launching from enemy territory. Charged with the task of investigating this, Hey concluded that whereas the V2 rockets actually produced echoes on the tube they were only a fraction of the transient echoes seen by the operators which probably had an ionospheric origin and might be associated with meteors.

My own wartime associations were with the R.A.F. and Air Ministry, and not with the Army, and I knew nothing of Hey's secret Army report until well into 1946 when I was already concluding that these echoes were not from cosmic ray showers but from meteors. The pre-war ionospheric literature was plentifully scattered with proposals that the trail of electrons left by a meteor burning up 100 km in the atmosphere scattered radio waves back to earth to give a detectable short-lived echo. But how could we obtain definite proof? Who were the authorities on meteors or shooting stars? The astronomical text-books contained little about the phenomenon.

Blackett's laboratory in the spring and early summer of 1946 was already gaining its characteristic cosmopolitan atmosphere and amongst his young visiting scientists was a delightful Norwegian, Nicolai Herlofson,[1] who had survived some hectic wartime experiences as a meteorologist. There is no connection between meteors and meteorology, but a meteorologist looks at the skies and should therefore know something about shooting stars. I don't think that Herlofson knew a tremendous amount about the subject when I asked him, but he did have one vital piece of information. It was that on the whole the study of meteors did not form a great part of the activities of professional astronomers because it was considered a waste of telescope time to look at phenomena which could be studied by the unaided eye. Consequently the investigation of meteors was largely in the hands of amateurs. Herlofson also knew that in England there was an active group of amateur meteor observers organized in a section of the British Astronomical Association. So Herlofson was dispatched to extract information from the director of this section, J. P. M. Prentice.

Manning Prentice was, and is, a solicitor in Suffolk. It was a revelation to me to find out how ardent he and his section were in observing

[1] Subsequently Herlofson was appointed to a professorship in the Royal Institute of Technology, Stockholm.

meteors far into the night after their daily work had finished. The published works of these people and their predecessors were priceless contributions to astronomy. Their enthusiasm was terrific and before the summer of 1946 had ended the lanes of Cheshire were graced by a new sight—the arrival of Manning Prentice in an open Morris with the back seat piled with celestial globes, star atlases, and other paraphernalia which he used for observing. A deck chair was precariously strapped on to the back hood and a flying suit was thrown over some of the instruments. We were to learn quickly that the deck chair and the flying suit were the key items in this collection.

For the first few months of Jodrell I was a lone figure with Alf and Frank to help me start the diesel. But I had Blackett's permission to recruit whom I could of my wartime associates either to work in his laboratory or to help me at Jodrell. By the time Manning Prentice arrived I had two or three colleagues with me in the trailers. The most important of these scientifically was John A. Clegg. There were two others, B and P, who lived in the same digs. B insisted on adorning the back of his bicycle cape with an identification sign so that on one dark night when cycling home from Jodrell he was mistaken for an escaped prisoner. That night was memorable because P, who was to take over the watch from me at 10 p.m., never arrived in the trailer. I was assured by his landlady that he had departed in his car and so I sleepily continued making records until at dawn I found that P had indeed arrived at Jodrell but had unfortunately fallen asleep before getting out of the car.

J. A. Clegg was the first of the many brilliant young men who were, in the course of time, to join me at Jodrell.[1] But few of these later ones who enjoyed the benefits of heated laboratories, canteen, office and supporting staff can have much idea of what scientific research meant at Jodrell Bank in 1946. To start a night's observation Clegg and I often began by wading in gumboots through two fields to our trailers and wrestling with the diesel, unprotected except by a flimsy canvas on its own trolley.

At some stage early in 1946 I had borrowed Moston and his tractor to tow the trailer containing the receiving part of the equipment from its original position to the third and most distant of the three fields owned by the University. This was done for technical reasons, but made

[1] Dr. J. A. Clegg remained with me at Jodrell until 1951 when he became interested in the new developments in rock magnetism then being stimulated by Blackett in the physics dept. of the University. After a short period as Professor of Physics at the University of Ibadan, Nigeria, he returned to Imperial College, London, and rejoined Blackett's staff there.

operations even more difficult, because the diesel, transmitter, and receiver were now all separated by hundreds of yards of Cheshire mud. We therefore proceeded tc make a road. I've made many roads at Jodrell since 1946—or rather I've arranged for them to be made by the people whose job it is to make roads. Then things were different, and when I write that we made a road, I mean that Clegg and I made a road ourselves. We started with spades, but since we wanted it to go from the main road to our trailers—a quarter of a mile—we found it would take a long time, so we again pressed Moston's tractor into service. Fortunately it had a scoop with which we carved out two tracks about a foot deep. We then ordered many, many lorry-loads of stone from a nearby quarry and filled the tracks by hand. With reasonable care we could drive a car with its wheels on these two rough tracks without sinking it in the adjoining mud, and so we were able to cut out at least some of the tiring night walks in gumboots.

We were in that kind of state when Manning Prentice arrived on the scene. Our idea was quite simple. If our echoes were from meteors then the echo should appear on the tube at the same time as the visible trail of the meteor appeared in the sky. We soon learnt from Manning Prentice that meteors were irregular in their rate of occurrence. There was always a sporadic background with a diurnal variation in rate, which at 6 a.m. in the morning, when the rate was a maximum, should give 5 or 10 meteor trails per hour visible to the eye under reasonably clear sky conditions. However, there were far more exciting events known as meteor showers, in which for a few nights the earth crossed through a relatively dense region of debris in space, and then the visual rate might reach 50 to 100 per hour. These meteor showers are called after the constellation in which their radiant point appears to lie. The Perseids in late July and August were according to Prentice a 'reliable and regular shower' giving hourly rates of 50 or 60 at maximum. The Perseids were convenient for our first joint venture.

We then discovered that the deck chair in Prentice's car was not for leisure but for observing. With the chair as near horizontal as possible Prentice would settle himself with a piece of string, a dimmed torch, and a writing board. When the meteor streaked across the sky he would raise the stretched string at arm's length along the line of the transient meteor and read off with comparative leisure the stars which defined the beginning and end points of the meteor's track. His writing board soon became covered with symbols which, in the beginning, were mere hieroglyphics to us. We arranged that Prentice should shout when a meteor appeared so that we, gazing at the cathode ray tube in the trailer, could establish any correlations between the echoes and the meteors.

Experiments in science are rarely decisive and this one was not. It was certainly the case that there were many occasions in which Prentice's shout coincided with the appearance of the echo on the tube. However, there were many shouts without echoes, many echoes without shouts, and strangest of all many shouts which seemed to be followed after some seconds by an echo. The echoes themselves were a mixed bag. Some so short lived that they were scarcely visible and others of many seconds' duration, generally showing violent fluctuations in strength. It was to be many years before we understood these complex phenomena.

The immediate task was to convince ourselves and others that our echoes were meteoric in origin and the Perseid episode gave much valuable information[1] but did not prove conclusively the complete correlation of the echoes with meteors. Now we were blessed by an extraordinary piece of good fortune. After the Perseids, Prentice said that he would return in mid November because the Leonids were the next regular shower worthy of investigation, although he said that on the night of 9–10 October he and his observers would be watching the sky intently because there was a chance that the earth would pass through the tail of a periodic comet and there might be an enormous but short-lived meteor shower. Thirteen years previously the earth had crossed the orbit of this Giacobini–Zinner comet near the head of the comet and there had been a spectacular meteor shower. He refused to commit himself on the chance of a recurrence because even small perturbations of the comet would deflect its accompanying debris out of the earth's orbit.

A look at the history of the comet, which was discovered only at the beginning of the century, and of the subsequent Giacobinid meteor showers associated with it, showed us that if there had been no perturbations then the shower of 9–10 October might indeed be spectacular. In our anxiety not to miss anything we arranged a continuous watch of the cathode ray tube for forty-eight hours around the predicted time of occurrence. We had optimistically fitted up a cine-camera to photograph the tube should the echoes be so numerous as to defy counting.

The hours dragged on to midnight, 9 October. We were seeing two echoes an hour which was quite a normal background rate. The sky was clear but nothing visual was evident. Then in a flash, it seemed, everything was transformed. Just after midnight our echo rate began to rise dramatically and simultaneously meteors streaked across the sky.

[1] The results of this early work were published in the Monthly Notices of the Royal Astronomical Society, vol. 107, p. 155, 1947.

The rates of echoes and visible meteors continued to soar. Soon they were coming so fast that we were unable to write down any details and by 2.30 a.m. there were so many that we could not even count them. We were glad of the cine-camera. By 3 a.m. the sky was streaked with trails and looked like the drawings of the great meteor showers of the eighteenth century which we had always thought to be imaginative. That was the peak. Our echo rate, which had been nearly a thousand an hour for a few minutes at 3.10 a.m., decreased and by 6 a.m. all was normal, except that those of us who were awake that night were in a state of great excitement. Moreover, term had begun, and both Clegg and I had to give lecture courses in Manchester at 10 a.m.

The dramatic correlation of the echo rate with the meteor rate in the sky finally convinced everyone that the radio echoes were clearly associated with the ionized meteor trails, although it was equally obvious that there were many peculiarities to be investigated.[1] It was fortunate that Patrick Blackett came out for some of the night. Not that I think he ever had doubts that we were on to a good thing, but his anxiety to help me was increased to the extent that he began to lobby the Vice-Chancellor and others on our behalf.

During 1946, aided and abetted by Clegg, I acquired much more apparatus. I had left many friends in the Air Ministry and a brisk exchange of personal letters with a certain Air Marshal quickly released for me several trailers of radar equipment, for which I think I paid a nominal sum of £10, although to buy it in the commercial market would probably have cost half a million. Clegg enjoyed collecting it. One of the items was a large cabin packed with electronic equipment, built on to a prime mover which was commonly known in the service as a 'Park Royal'. One day Clegg joyfully appeared with this, having driven it from somewhere in the south, negotiated our tracks, and proceeded across the field to a point which we had assigned to it. The Park Royal never reached that point. Half way there the mud proved too much and it stuck. It remained in that position for many a day and determined the siting of several of our aerials, and eventually of our first permanent buildings. The building which, some years later, was erected near to this trailer to take over its apparatus was instantly known as the Park Royal and remains so today, although I very much doubt if more than a handful of people at Jodrell today know the origin of its name! It was from this beloved Park Royal, stuck in the mud, that we observed the fantastic Giacobinid meteor shower.

[1] These observations of the Giacobinid meteors were published in the Monthly Notices of the Royal Astronomical Society, vol. 107, p. 164, 1947.

I extracted apparatus from the Army and Navy as well as the Air Force, but I did not have the same kind of contact in the higher echelons of these services as in the Air Ministry and things were more difficult. We did, however, manage to get an Army searchlight which we badly needed—not for its light, but because it was an excellent turning mechanism on which to mount an aerial. Eventually we signed forms to have it on loan, but within a few weeks it was obvious that this particular device would never do service with the Army again. Clegg was an expert on aerials, and on this searchlight mount he built a magnificent broadside array which could be turned to any part of the sky. The possibility of being able to direct the aerial was important because we found that most of the echoes from the meteor trails were obtained by 'specular reflection'—that is, the echo was obtained only when the ionized column of the trail passed through the aerial beam nearly at right angles to the direction in which it was pointing. This aerial was a few yards from the Park Royal and Clegg later used it and the property of specular reflection to devise a new and fundamentally important method of determining the radiant points of meteors—the direction in space from which they apparently diverged. The Giacobinid shower presented us with a marvellous opportunity for a conclusive demonstration of this property of meteor reflection. While the echo rate was increasing the aerial had been maintained in a direction at right angles to the position of the radiant point in Draco as it moved across the sky. Near the maximum, when the echo rate was many hundreds per hour, we turned the aerial so that it was directed *at* the radiant point. The sky remained brilliant with meteor trails, but the echoes on our cathode ray tube disappeared almost completely.

With the correlation between the radio echoes and meteors finally established, Manning Prentice's enthusiasm was redoubled. At last he could find out what the meteors were doing on cloudy and moonlit nights, and for our part we badly needed his visual correlations to help us unravel the complexities of the echoes. So it was that Prentice became a frequent visitor and that winter we understood the reason for the flying suit. A deck chair at 4 a.m. on a snow-covered field near the time of maximum of the Geminid meteors in mid December is a severe test of a man's observing ability. It was a great pleasure to me some years later when the University of Manchester awarded Prentice the honorary degree of M.Sc. As an amateur he had given of his spare time (and I suspect of his business time) unsparingly to help us and his influence was cardinal to our early development and education in astronomy.

3

The fairground

In 1947 the field at Jodrell bore a strange appearance: a considerable number of trailers still in their camouflage paint, several portable diesels, including some bigger ones which were even harder to start, and various aerials—all of which led the locals and some visitors to describe us as the fairground. I mentioned earlier that after the Giacobinids Blackett thought he would introduce the idea to the Vice-Chancellor and other controlling officials of the University that our activities at Jodrell had gone somewhat beyond the idea of a temporary trailer in a field.

The Vice-Chancellor of the University of Manchester was then Sir John Stopford, F.R.S. (later Lord Stopford of Fallowfield), an eminent medical scientist, an outstanding Vice-Chancellor, moreover a man beloved by all and a person of unerring surety of judgement. The significance of this visit as far as our future was concerned was not lost on me. We had reached the stage when we urgently wanted not merely an official blessing to continue, but some form of permanent accommodation for the precious apparatus in our leaking trailers, more money for our research, and more staff carrying out research twenty-five miles away from the University. Vital matters of University policy were beginning to be involved.

Blackett arrived with the Vice-Chancellor on a golden June morning. It was our good fortune that, in May, we had realized the significance of an extraordinary high echo rate which had been present the previous year. It had again returned and Clegg, by using the 'searchlight aerial', had located the radiant point of a vast region of debris through which the earth was moving, providing meteor showers in daylight quite invisible and unknown to the human eye. We ourselves were highly excited because the rate continued to increase as the summer advanced and we were determining a sequence of these summer daytime meteor showers. When Blackett had asked me when to bring the V.-C. I begged him to do so early in the morning when we could display these meteor streams and not in the afternoon when the radiants would have set. So it happened that the occasion was perfect and when the V.-C. entered the darkness of our trailer from the brilliant June sunshine he was confronted with a screen full of these echoes. It was the first of many critical occasions when the Vice-Chancellor held our future in his hands, and although we were to cause him much trouble and anxiety I know that he never for one second regretted his initial and spontaneous enthusiasm and approval.

Most of our time in those early years was occupied with the investigation of these echoes, and as the staff slowly increased so our techniques improved. The visual watch of the tube gradually became replaced by automatic photography and we soon learnt that important information was to be obtained about the physical properties of the high atmosphere and ionosphere in addition to the study of the astronomy of the meteors. For example, we found out that the variation in amplitude of the echoes was caused by the contortion of the trail under the influence of the ionospheric winds, and in the years which followed many papers were published from Jodrell on these physical aspects of our work. Little did we realize then that ten years later this equipment and our techniques were to serve us in a radically different kind of experiment.

I think that perhaps the most important of our developments at that time was carried out by a young research student who had just joined me, John G. Davies, and a visitor from New Zealand, C. D. Ellyett. This development was based on an idea by Herlofson who, although not a nominal member of my staff, remained our close friend and adviser. The stimulus arose because of the urgent need to find some reliable method of measuring the velocity with which the meteors entered our atmosphere. As we became more familiar with the literature on the subject of meteors we realized that there was a bitter dispute about the nature of the sporadic meteors—that is, those meteors which can be seen on any night as distinct from those like the Perseids and the Geminids which are concentrated in showers. In the case of these shower meteors it was obvious that they were closely associated with comets, although even in that case there was no agreement as to whether the meteor particles were the debris of the comets or were otherwise genetically associated with them. The orbits of the comets and hence of these shower meteors were well known: they were moving in elliptical orbits around the sun and were confined to the solar system.

There was no such certainty about the sporadic meteors; and many people believing the evidence from two different sources of measurements by reliable astronomers firmly held the opinion that these meteors were not confined to the solar system but were visitors from interstellar space, temporarily captured by the sun and moving in a hyperbolic orbit around it. Astronomically, the issue was of some importance, since if the sporadic meteors were from interstellar space then they would give us significant information about the size and distribution of the interstellar dust.

It had been found impossible to settle this dispute by the normal methods of observation. In order to deduce the orbit in which a meteor is moving in space it is necessary to measure both its velocity and direction with considerable accuracy. For example, the limiting velocity of a particle moving around the sun in a closed or elliptical orbit at the earth's distance is 42·2 km/sec. If its velocity is greater than this then its orbit is open and hyperbolic and it must be a visitor from space. This is known as the heliocentric velocity: the velocity around the sun. But when the meteor enters the earth's atmosphere it does so with a velocity compounded of this heliocentric velocity around the sun and the earth's own orbital velocity. The earth's orbital velocity is 29·8 km/sec. and hence the observed geocentric velocity of a meteor which is actually moving at 42·2 km/sec. around the sun can vary between 12·4 km/sec. (42·2 − 29·8) and 72 km/sec. (42·2 + 29·8) depending on whether it enters the earth's atmosphere on the evening or morning side. Apart from the computation of the precise orbit it is obvious that under no circumstances can the observed velocity of a meteor moving in an ellipse around the sun exceed 72 km/sec. If it does so then its orbit is hyperbolic and it is not confined to the solar system.

At the time when we became acquainted with this problem the subject was in disarray. The visual measurement of such high velocities from transient appearances of the meteor trails was too approximate to enable these fine distinctions to be made and the accurate photographic measurements which were later made by new types of camera at the Smithsonian Astrophysical Observatory still lay some years in the future. Stimulated by this situation, Herlofson suggested that the radio techniques could provide a precise answer to the problem if apparatus could be developed to record the amplitude of the echo from the trail of the meteor every millisecond or so as it passed through the beam of the aerial. He suggested that in this case we should see the Fresnel diffraction pattern and since we also measured the range accurately, then the velocity would be known with precision. In the apparatus developed by Davies and Ellyett to observe this phenomenon the first responses from the echo were made to trigger the camera mechanism and they obtained photographs of the echo amplitude every thousandth of a second.

The technique was brilliantly successful. Meteor velocities could now be measured with unerring accuracy and, so I thought, the problem of the origin of the sporadic meteors was solved. In a flurry of pleasure and excitement I sent the results to the chief adversaries in the contest. Dr. Öpik[1] replied, 'Your results confirm my measurements that the meteors are from interstellar space'. Dr. Porter[2] replied, 'Your results leave no doubt that the meteors are contained within the solar system'. Professor Fred L. Whipple[3] of Harvard, one of the great authorities on meteor phenomena whose ground at that time was neutral in the contest, replied, 'Your results are indecisive but give promise of a valuable contribution to the subject'. How wise he was! Many years later Davies and I carried out a long series of measurements with the assistance of Mary Almond, and in a number of papers published in the Monthly Notices of the R.A.S.[4] we showed that the hyperbolic theory was untenable and that the sporadic meteors, like the shower meteors, were moving in closed orbits in the solar system. Simultaneously different kinds of radio techniques in Canada produced similar results, and the new Schmidt cameras of the Smithsonian Observatory gave the same answer. This work killed the theory of the interstellar origin of the meteors (a pity I think; it would have been more interesting if they had been interstellar!); but it did not solve the problem of the nature and origin of the sporadic meteors in the solar system. Ten years after this episode Davies devised an even more sophisticated technique by which he could measure the exact orbit of a single meteor. The mass of data which he obtained showed that many of the sporadic meteors (far from being in hyperbolic orbits) were actually moving in elliptical orbits of short period. The reason for the peculiar distribution of these short-period orbits in the solar system is a problem without solution today.

Since I had originally gone to Jodrell Bank to look for radio echoes from cosmic ray showers, it was natural that we should spend a good deal of time investigating the echoes which did turn up. We have seen that there were large numbers of them and that over many years we came to understand a good deal about their behaviour and about the meteors which caused them. The appearances of these echoes on the cathode ray tube were varied and we were used to seeing all types—some occasionally lasting for several seconds. One night early in August 1947 when we were studying the Perseids a new phenomenon

[1] Dr. E. J. Öpik of Armagh Observatory.

[2] Dr. J. G. Porter, at that time of the Nautical Almanac Office, Royal Greenwich Observatory, and director of the Computing Section of the British Astronomical Association.

[3] Prof. F. L. Whipple, the director of the Smithsonian Astrophysical Observatory and Professor of Astronomy at Harvard University.

[4] vol. 111, p. 585, 1951; vol. 112, p. 21, 1952; vol. 112, p. 399, 1952; vol. 113, p. 411, 1953.

appeared on our tubes. The meteor echoes were there, precisely defined in range as usual. Suddenly a diffuse echo appeared on our tube extending over a large part of the time base (that is in range) and having the appearance of much frothing and bubbling as though the whole was composed of a multitude of individual scattering centres. Moreover, this phenomenon persisted for seconds and then for minutes, moving to and fro along the time base. We were astonished. Fortunately the night was clear and outside the cabin we could see great auroral streamers stretching from the northern horizon nearly to our zenith. Frantically we did whatever measurements we could. We had two equipments working on different wavelengths—including the steerable aerial—and we were able to establish without doubt that our echoes came from the aurora. By good fortune we had opened another line of research which for some years provided a fruitful subject for many of my students and gave important information about the nature of the aurora borealis. Nowadays radio studies of the aurora are widespread in many laboratories, particularly those at high latitudes where the phenomena are of frequent occurrence. But that August night in 1947 contained something new, and like the sky outside it was spectacular and memorable.

Echoes on the cathode ray tube were not the only phenomena to impress themselves on our apparatus. Towards the end of July 1946, when we were about to enter on our combined programme with Manning Prentice, the noise level on the tube increased to such an extent that our echoes were obliterated. I thought the apparatus had become faulty, but this was not so. The noise increase was occasioned by a gigantic solar flare on the sun. Today there is a vast literature on radio emission from the sun, but in the summer of 1946 very little was known about the effect. Some years before the war Grote Reber had searched for radio emission from the sun in the metre waveband but had failed to detect any. I have already mentioned J. S. Hey's secret paper on the echoes from meteors which were confused with V2's. A year or so previously he had been called in to investigate a suspected case of jamming on the A.A. radars. In February 1942 the operators had found the noise level on their tubes increasing to such an extent that the echoes from aircraft were obliterated. They suspected jamming by the enemy. Hey concluded that the rise in noise was associated with the great solar flare active at that time. I cannot now remember whether I knew of the existence of Hey's secret paper at the time of our observation, but the July 1946 event was observed by several other workers besides ourselves and left no doubt about the violence of the radio emissions from the solar flare.

4

The first large aerials

ALTHOUGH the echoes on the cathode ray tube quickly led us deep into the study of meteors it was some years before I abandoned the original intention of searching for radar echoes from cosmic ray showers. Indeed the development of our ideas in 1946 and 1947 as to how this should be done eventually led finally and irretrievably to our present destiny.

I have mentioned that Clegg was an expert on aerials and that he built the large broadside array on the searchlight mount. My calculations on the cosmic ray problem indicated that success would demand much greater sensitivity than we possessed and that the only way to get it at that stage of our technique was to build a large aerial. Clegg and I therefore started building a huge tower from hired scaffolding tubes on which to mount the array. In retrospect I am astonished at the manner in which we gaily started not inconsiderable engineering constructions without either of us having any professional competence to do so. The proposed tower obviously needed fixing to the ground and since we wanted it to be at least a hundred feet high and as wide we decided to anchor the first scaffolding poles in concrete blocks. We had no cement, sand, stones, or concrete mixer. We soon collected enough of the ingredients but a mixer was quite beyond our resources so we mixed the concrete by spade work.

I think that on the first day of operations I was away lecturing in the University and left Clegg with instructions to dig eight holes two feet wide and fill them with concrete. When I returned home in the evening my wife said that late in the afternoon she had received an agitated phone call from Clegg to inquire if she thought I realized how long it took one man to perform such a task! Part of the tower was eventually built—I have photographs of it now—but we never reached the stage of constructing an aerial on its framework.

One reason for the abandonment of the tower broadside aerial was that we had a much better idea. We would build a giant wire paraboloid on the ground. This we rightly held would be much better because it would be easy to change the working wavelength, and easier to feed it with a powerful transmitter. Once more we worked out our plans on the basis of our own intuition. The size of the bowl was determined by the distance between the place where the Park Royal trailer had stuck in the mud and the hedge of our field. Taking into account the straining wires we decided that the bowl could be 218 ft. in diameter. Our idea was to build the paraboloid of wire so that it looked vertically. So the next problem was the height of the edge of the bowl at its periphery. This was determined by the necessity that we must easily be able to reach it on a ladder; so it was fixed at 24 ft. The diameter and depth of the bowl determined its focal length: 126 ft.! There our amateurish ingenuity failed us. Clegg knew about wooden towers of about that height that were used on coastal radars (C.H.) during the war. So there were more telephone calls to the Air Ministry and eventually Clegg arrived with a lorry-load of wooden struts from which the tower had to be built. We actually got the concrete work and base section of this tower built before the end of 1946, I think. Then we had to suspend work for the winter.

I have no record of why we abandoned the wood tower at that stage. It could be that discretion intervened on what would undoubtedly have been a difficult and dangerous task for two inexperienced people. Also I suspect that our financial situation was becoming increasingly favourable. In any case the base of the wooden tower was removed and in the early spring contractors put in new foundations for a 126-ft. high steel tube, pivoted at the base and held by guys. The idea was that we would gain access to the focus by a bos'n's chair. I never enjoyed that perilous undertaking. Fortunately the steel tube was hinged at the base and by manipulating the steel guys it was possible to lower it so that easier access to the aerial at the focus could be achieved.

The construction of the main part of the paraboloidal bowl was carried out during the spring and summer of 1947. By that time Clegg and I were no longer lone hands, but the few people who were with us certainly learnt how to mix concrete. The first task was to erect the twenty-four perimeter scaffolding tubes, fixed in concrete blocks. The main framework of the bowl was then formed by $\frac{3}{8}$ in. steel wires radiating from the centre to the tops of these posts and strained to the ground in more concrete outside the perimeter. These wires were held to the ground by more vertical posts along the radius. On this framework we wound miles and miles of thin 16-gauge wire to form the reflecting screen. In the outer section we drove a lorry around the inside of the bowl and by standing on a tall platform on the lorry we were able to loop the wire over its supports with the lorry in motion. The tying of this wire to the supports with more thin wire was a tedious job and those which could fairly easily be reached from ground level we left to our wives and children.

This large transit telescope, as it became known, was a great success. Early in the autumn of 1947 we tried some experiments with it to detect the cosmic ray showers but by that time the new instrument was proving to be so valuable in other fields of work that we eventually gave up the cosmic ray experiment.[1]

Although we did nearly all the constructional work ourselves, the material for this telescope, particularly the 126-ft. steel mast, cost money—about £1,000 I think; not much by our standards today but far more than I had at my disposal for research in 1947 from University sources. It may therefore be wondered how we afforded this amount. The answer is that Blackett had introduced me to the machinery of the D.S.I.R. (Department of Scientific and Industrial Research). One Saturday morning during our early days I drove the D.S.I.R. official in charge of university grants to Jodrell Bank with no expectation of any favourable response. To my surprise after I had shown him some echoes in the trailer and explained what was going on, he said, 'This is exactly the kind of research of timeliness and promise which the Department likes to support'. I asked, I think, for £1,000 to complete the building of the large aerial and got it without hesitation.

In the years which followed Jodrell changed its face rapidly and it is not my intention to describe in any chronological form the details of our growth once those unforgettable early years were past. Suffice it to say that the influence of the Vice-Chancellor's interest soon materialized in the form of permanent buildings which replaced the trailers. We acquired a technician, labourers, and an engineer to run our diesels. The old wooden hut was transformed to a small canteen. We built a small dormitory. Inevitably filing cabinets appeared, and the young lady[2] who had so nobly done my office work in a cold trailer with a typewriter surrounded by wire and electronic gadgetry, acquired an office.

The faithful technician and engineer remained with me throughout all the events described in this book. Edward Taylor (Ted), who arrived in 1947 and became Chief Technician and head of our Workshop until he retired in 1966, still does part-time work in the Visitors' Building. The engineer A. W. Smith (Fred) came in 1948 and has cheerfully and resourcefully borne the vast increase in his responsibilities over the years.

[1] By a strange twist of fate a successful experiment to detect the emission of pulses of radio energy from cosmic ray showers was carried out at Jodrell Bank in 1964 by Prof. Graham Smith and his colleagues. As far as I am aware the radio echo experiment which I originally proposed has never been revived.

[2] Miss Ella Ryder, subsequently Mrs. Ella Bradley, who still works part of the day at Jodrell.

5

Radio astronomy

At the time of which I am writing I do not think that the term 'radio astronomy' had been coined. Several small groups of post-war researchers like ourselves were at that moment working around the fringes of an entirely new subject, which under that name was soon to assume major importance in the development of science, and of astronomy in particular. In order to place the work of the transit telescope and the evolution of its great successor in perspective it is necessary to pause in order to consider the state of our knowledge of the universe in the period of about 1947–49.

Until that time almost our entire knowledge of outer space was achieved because man has evolved with eyes which are sensitive to radiations in a small region of the spectrum from the violet to the red. Radiation from the stars in this part of the spectrum penetrates the earth's atmosphere without appreciable absorption. The region is a window through which we can look into the universe. At wavelengths shorter than the violet or longer than the red, any radiations are absorbed by the dust or water vapour in the atmosphere. It was well known from the fundamental laws of physics that the sun and the stars give out most of their energy in this visible part of the spectrum. Our inability to study the radiations from space beyond the red and violet was not therefore considered to be seriously detrimental to our study of the heavens, although of course it was recognized that this restriction of our view hindered our knowledge of many of the basic processes occurring in the sun and the stars.

Through this narrow window in the atmosphere man had been able to obtain a vast amount of data about the universe. The ancient idea that the earth was the centre of creation was shown to be untenable by Galileo with his 2¼-in. telescope 350 years ago. In the succeeding centuries ever larger telescopes were built which, because their mirrors collected more light, were able to penetrate even deeper into space. Nevertheless until the period 1920–30 it was still thought that the sun and the solar system were at the centre of the universe, a universe defined by the stars which formed the Milky Way.

Our modern ideas did not emerge until the 100-in. telescope on Mt. Wilson came into use after the First World War. With this instrument Harlow Shapley found that the stars in the Milky Way were not symmetrically disposed about the sun, but that they were in the shape of a flattened disc. Our present figures for the dimensions of the Milky Way are somewhat different from those initially derived by Shapley but his general picture remains intact. The 100,000 million stars which comprise the Milky Way extend for 100,000 light years. They are contained in great spiral arms emerging from the central nucleus of the galaxy. In these central regions the disc is about 20,000 light years thick. The sun and its family of planets is far out in one of the spiral arms 30,000 light years from the nucleus. The entire system is rotating and at the sun's distance from the centre one rotation takes about 200 million years.

This revelation of the insignificance of our location amongst the stars of the Milky Way was followed by even more remarkable discoveries. The existence of diffuse nebulous patches in the sky had been known for centuries. Kant had suggested that they might be 'island universes' outside the Milky Way, and Herschel inclined to this view, but the decisive evidence was first obtained by Edwin Hubble with the 100-in. telescope. He showed that some of these nebulae were, indeed, great star systems or galaxies far removed from the Milky Way. The nearest of these, similar in size and structure to the Milky Way, is the great spiral M31 in Andromeda. By identifying certain types of star in this nebula Hubble placed the distance at 700,000 light years, but we believe now that its distance is 2 million light years and that the nebula contains 100,000 million stars in a spiral structure with dimensions similar to that of the Milky Way.

Large numbers of these extragalactic nebulae were soon recognized and it was found that their light was reddened, the spectral lines of identifiable elements being shifted towards the red end of the spectrum. Hubble interpreted this shift as a doppler effect arising from the motion of the nebula away from us. He found that the extent of this shift and hence the velocity of the recession of the nebula increased linearly with distance out to the furthermost limits of penetration of the telescopes.

At the period of which I am writing the 200-in. Hale telescope on Mt. Palomar had just come into use and this was still further extending our knowledge of these distant extragalactic nebulae. At that time the 200-in. had photographed clusters of galaxies to distances of about 2,000 million light years whose velocity of recession was 38,000 miles per second, that is 20 per cent. of the velocity of light. Within this vast region of space and time it was estimated that there were probably several million millions of extragalactic nebulae. These nebulae were of diverse structure and Hubble suggested an evolutionary sequence beginning with almost structureless elliptical or spheroidal galaxies evolving to various types of nebulae showing pronounced structure such as the spiral formation of the Milky Way.

Faced with these revelations of the optical telescopes there was little reason to believe that significant contributions to our knowledge of the universe could come from other parts of the electromagnetic spectrum. This belief existed although many years previously other radiations from space outside the visible spectrum had been discovered. In 1931 Karl Jansky of the Bell Telephone Laboratories was investigating atmospherics as a source of interference in long-distance radio communication. His apparatus worked in the radio wave part of the spectrum on a wavelength of several metres, and he noticed that, even when there were no obvious sources of atmospherics such as thunderstorms, a residual noise persisted in his receiving apparatus. He then made the simple but elegant observation that the strength of this noise varied throughout the day and that the period of its diurnal variation was 23 hours 56 minutes. This is a sidereal day—the period of the earth's rotation with respect to the stars—and from this Jansky was able to conclude unambiguously that the source of the noise which he was receiving must lie, not only outside the earth, but outside the solar system. He referred to the phenomenon as cosmic static, and decided that the stars must be emitting energy in the radio wave part of the spectrum as well as the light by which we see them.

The existence of this other transparency or window in the atmosphere had been known before Jansky's work but no one thought it would be of significance to astronomy. The gap exists from wavelengths of about a centimetre to several metres. The lower cut-off is caused by absorption in the water vapour of the earth's atmosphere. The upper cut-off at wavelengths of 15 to 20 metres is caused by reflection and absorption in the earth's ionosphere—the region of electrons which surround the earth at heights of 100 to 400 km. The wavelength of this cut-off is variable and ill defined: it depends on the angle of incidence of the radiation and on the condition of the ionosphere which is a complex function of time of day, season, sunspot cycle, and position on the earth's surface.

It is a most extraordinary affair that no one took any notice of Jansky's discovery. Indeed it is recorded that his laboratory asked him to proceed with the investigations on atmospherics. Fortunately he published his observations and some years later an amateur investigator, Grote Reber, made further investigations. Reber built a 30-ft. paraboloidal bowl which he could steer to different parts of the sky. He did this in the garden of his house at Wheaton, Illinois. This historic instrument, the first of the paraboloid types of radio telescope which are so familiar today, has been preserved and re-erected as a museum piece at the entrance to the United States National Radio Observatory at Green Bank, West Virginia.

Reber used a much shorter wavelength than Jansky (about 2m) and because of this and the size of his aerial he had much more directivity. He was able to confirm Jansky's conclusions and he made a rough map of the distribution of the strength of the radio emissions from different parts of the sky. He found that the radio signals were strongest when his aerial was receiving from the direction of the Milky Way where most stars were concentrated and that they were weakest from the parts where there were few stars. In fact, he discovered that the isophotes or contours of strength of the radio emission resembled the pattern of the distribution of the ordinary stars in the Milky Way. This seemed to confirm Jansky's idea that the stars were emitting the radio waves. However, when Reber pointed his aerial at the conspicuous stars in the sky he failed to detect any radio waves. He failed also to detect radio waves from the sun (although we know now that this was merely because he was observing at a minimum period of solar activity). He concluded, therefore, that the stars could not be responsible for the radio waves and that the interstellar hydrogen gas between the stars must be the source of the radio emissions. He described a process known as a free-free transition, in which an electron moves in the field of a proton without capture, as a possible mechanism. Reber's idea was to suffer many vicissitudes in the years to come but today we know that he was partly correct. When one remembers that Reber was a lone hand working in his spare time his achievement stands out as altogether remarkable.

Reber's work was published in an American scientific journal during the war and his observations, together with those of Jansky, constituted almost our total knowledge of the radio waves from space when the war ended. At that time I knew about these observations in a circuitous manner and had no conception that their further pursuit was to occupy so much of my post-war career. I remember precisely the occasion on which I first heard about the existence of this cosmic static. We were engaged in a desperate attempt to increase the sensitivity of some of our airborne radars in order to detect enemy submarines and other land targets at greater range. There were three avenues open to us: increase the transmitter power, increase the size of the aerial, or improve the sensitivity of the receiver. The first was at the limit of current techniques, and in the case of the second we had already caused consternation amongst the aircraft designers by our demands for large aerials. The improvement of receiver sensitivity seemed the easiest and most obvious course. Alas! This easy optimism was punctured by a member of my group who respectfully informed us about the existence of cosmic static which must inevitably set a limit to the receiver sensitivity which could be realized in practice, and in due course he produced copies of the papers by Jansky and Reber.

The study of these radio waves from space was not the immediate incentive for Jodrell Bank, where we were occupied with the problem of the echoes. Immediately following the end of the war, however, J. S. Hey while still at the Army Operational Research Group used the receiving part of the A.A. radar equipment (the G.L. set) to make further experiments on the extraterrestrial radio emissions. He made one observation of great importance: that from the direction of the constellation of Cygnus the radio emissions varied in strength with a short time period. He argued that this implied the existence of a localized body or star-like object in that direction which was emitting the radio waves. He was correct, although later it transpired that by a strange irony the argument through which he deduced the existence of the localized source was incorrect. (The variations in signal strength are not caused by variations in the strength of the waves emitted but they arise by a diffraction mechanism when the radio waves pass through the earth's ionosphere. The effect is analogous to the twinkling of an ordinary star.)

Hey's observations were made by his group in the A.O.R.G., and similarly in Australia a group under the late Dr. J. L. Pawsey, in the Division of Radiophysics of the Council for Scientific and Industrial Research, turned their radar apparatus to the study of the emissions from the sun and from space. In England, Martin Ryle returned to the Cavendish Laboratory in Cambridge and started his own group which soon began to obtain results of vital importance to the development of the subject. The Australian group verified Hey's suggestion that a localized source of radio emission existed in Cygnus by using an instrument known as an interferometer, and almost simultaneously Ryle in Cambridge with a different type of interferometer found another intense localized source of emission in Cassiopeia.

In the Sydney group John G. Bolton had discovered another source in Taurus and his suggestion that this source was related to the Crab Nebula (the remnants of the supernova of 1054 A.D.) was soon generally accepted, but no one had any idea at that time of the identity of the sources in Cassiopeia and Cygnus. Even in a short time, it had at least become clear that Reber's theory that the radio waves were generated in the interstellar hydrogen gas could not be the whole story.

6

Dreams

On wavelengths of a few metres the transit telescope had a beamwidth at least six or seven times smaller than any other instrument which had so far been used in these studies. The early interferometers in Sydney and Cambridge had, of course, greater ability to resolve point sources but they had only the power gain of the small elements of which they were composed and not the full gain appropriate to the large aperture of this new telescope. Consequently the variation in the strength of the signals from the sky in the transit telescope stood out like the humps on a camel's back.

After a few days the constant repetition of this pattern almost became an irritant. If only we could shift the beam it would be possible to build up a map of the sky and not simply that of the emission from a narrow strip.[1] Moreover, there were other cogent reasons for the desire to be able to shift the beam. The tower aerial which we abandoned would have had a beam looking nearly horizontally, and Clegg and I were constantly discussing the optimum direction for the beam in the cosmic ray equipment. However, we really only had the option at that time of building it either looking horizontally (the tower) or vertically (the paraboloid) and (fortunately as it transpired) we had gone for the latter.

I cannot remember when or where I first summoned up enough courage to tell Patrick Blackett that I had visions beyond the transit telescope. It was probably soon after the instrument began to work in the spring of 1948. I was becoming more and more obsessed with the possibilities for all our interests—cosmic rays, meteors, aurorae, the moon and the planets (with which we were also dabbling), the radio waves from space, and the sun. However, I should have known better than to have had any doubts about the outcome of such a conversation. I explained to Blackett that I wanted a telescope like the transit one, of at least that size, but which we could steer to look at any part of the sky, and I told him what I thought we could do with it. Without a second's

[1] Some time later a strip of sky several degrees wide about the zenith was mapped with this telescope by the device of displacing the aerial laterally from the focal point.

hesitation he said that if I could find out some means of building such a device and the cost, he would back me to the hilt.

After the initial elation I felt like a man in the wilderness. The structures already existing at Jodrell were our amateur efforts which owed nothing to any engineering knowledge or advice. I had no idea of the procedures for finding the people who might be interested or of the methods of getting an estimate of the cost of a project like this. I was to learn the hard way.

To begin with I wrote to a few of the largest engineering firms in the country to explain what I had in mind. The replies were unanimous in one respect—that the firms were not interested. The reasons were either that they were too busy on work of 'national importance' or that such a project represented an impossible engineering undertaking. This was a depressing start and evidently was the wrong way to begin. At about that time a friend came to see me. He was an important member of a large engineering firm who had much to do with large optical telescopes. He was excited about the idea of steering a telescope of this size and immediately suggested various principles, such as the use of hydraulic rams, by which it might be done. Alas! he was quite certain that his own firm would not be interested in the project but he suggested another firm, which had a special department for unusual jobs, that he thought would be thrilled at the idea of tackling this.

This organization was, and is today, an important firm in the British engineering industry but my contacts with their special projects department brought us to the brink of disaster as far as the idea of the large telescope was concerned. After the initial introductions I faced the relevant people and explained what was needed: a steerable paraboloid at least as big as the transit telescope, preferably even larger. After some consideration they informed me of their interest and that they would be willing to carry out a preliminary investigation to find out if it was possible and give me a rough idea of the cost for the sum of £1,000. In due course I reported this to Blackett and asked him for the £1,000. In his great wisdom he said that he would ask the University Council to finance the investigation. In his view this would make the University officially associated with the project and would be an important point when we reached the stage of approaching the D.S.I.R. with the request for the full amount of money to build the instrument.

These formalities were completed and I began the discussions with the firm's special projects department. This department was in the vicinity of London, several hundred miles from the main works of the company. The head of this section had suffered severe war injuries and it is I think only generous, in retrospect, to bear in mind that this may have accounted for many of the subsequent happenings. There is no doubt that at the time of the preliminary discussions he was tremendously keen on the idea of the large telescope and asked for a few months in which to make a general arrangement drawing of his proposals and a model. In the early summer of 1950 everything was ready for me to take Blackett to see the results of our discussions.

At that time we had little idea of the cost of the enterprise on which we were proposing to embark; but we had of course a rough idea of what we felt it ought to cost. In fact the figure which had been talked about internally in the University was of the order of £60,000 to £70,000. The visit began badly. Blackett had reserved the train journey from London to discuss some points with me and we had arranged to meet on the train at Paddington. We missed. We both thought the other had lost the connection. In fact he was at one end of the train and I was at the other. We met on the arrival platform at the very end of the journey.

The tone of the whole visit had been set. I knew immediately that Blackett reacted unfavourably to the head of the projects department. The dream telescope which was 300 ft. in diameter was there in the shape of a beautiful model but we were quickly informed that such an instrument would cost at least half a million pounds. The special projects director then worsened the situation beyond repair by producing a stream of ideas for redesigning the device, particularly by suggesting various schemes for a polar axis mounting in which we had no faith. I think that by the end of the afternoon we had agreed to meet again in two months by which time there would be an estimate of a real price on which it might be possible to negotiate for a smaller version in the region of 200-ft. aperture.

Things were going badly. I had lost much ground with Blackett who clearly was not prepared to move further unless he was convinced not only of the soundness of the scheme but also about my own competence to deal with the issue. He was already pressing an argument that it would be better to start with a very much smaller and cheaper telescope. I argued that the state of the science demanded at least a 200-ft. aperture and preferably larger.

It was under these tensions that we arranged the next meeting with the director of the special projects department who had promised to fly to Ringway in a private plane and drive out to Jodrell Bank. The meeting was fixed for the morning of 27 June 1950 at 10. Blackett arrived some minutes ahead of time. The special projects head never turned up. At 10.30 we phoned Ringway who had no knowledge of the visitor or his plane. At 11 we phoned the special projects department but the number was unobtainable. At 11.30 we got through to the head office of the firm who regretted to inform us that two days previously they had closed their special projects department and that the person we were waiting for was no longer a member of their organization. The letter which they said we should have received informing us of this had not arrived neither did it at any time thereafter.

At that time Blackett was a very busy man. He had much on his mind and I knew without him telling me that he could ill afford to spare a morning at Jodrell Bank on a wild goose chase. I could only say that it was a very beautiful summer morning anyhow, at which Patrick said he hadn't noticed it but perhaps I was right on that point.

The devastating nature of this situation as far as my idea of a telescope was concerned needs no further comment. I am glad to say, however, that the University flatly refused to part with the £1,000 to the firm.

7

A new start

In the meantime the representative of the only engineering firm with whom we had contact at Jodrell Bank—the people who had supplied us with the steel tube for the transit telescope—came on a visit to see if it was behaving all right. I confided my troubles to him and asked if he had any advice. He replied that he thought I ought to discuss it with a good consulting engineer who would not be attached to any particular engineering firm. He added that he thought he knew just the man.

So it was that on 8 September 1949 I met H. C. Husband on the perimeter of the transit telescope at Jodrell Bank.

'What's your problem?' 'I want a telescope of at least this size mounted like that small 30-ft. paraboloid over there so that we can steer it to any part of the sky. I've now been trying for over a year to persuade someone to do it but I'm told it's impossible.'

Husband looked at the device for a few moments, gazed up at the focal point 126 ft. above ground, and then said calmly: 'Oh, I don't know. It should be easy—about the same problem as throwing a swing bridge over the Thames at Westminster.' I had emerged from the wilderness!

Husband soon produced a sketch of his proposal.[1] He knew the figure we had in mind and said that it was impossible to price the telescope until the details were gone into but he thought that our figure might not be unreasonable. At that time Sir Ben Lockspeiser[2] was the Secretary of the D.S.I.R. Years later he was chairman at a lecture which I gave to the Royal Society of Arts[3] and in his opening remarks he recalled:

Shortly after the war I was sitting in my office, when a young man came and asked me to let him have £1,000 in order that he might build a radio telescope. After he had explained about his idea, I realized that he was proposing a very promising new line of enquiry, and he got his £1,000.

Two years afterwards, the same young man walked into my office and said

[1] This sketch is dated January 1950.

[2] Sir Ben Lockspeiser, K.C.B., F.R.S., was Secretary of the D.S.I.R. from 1949 until his retirement in 1956. He was Chief Scientist to the Ministry of Supply 1946–49 and had previously held several important Government aeronautical research posts.

[3] *Journal Roy. Soc. Arts*, vol. CIII, p. 666, 1955.

he wanted £100,000 to build a new radio telescope. He pointed out that this would provide a new method for investigating the universe, and because it was independent of cloud and other climatic conditions which make observational astronomy in this country so difficult, he felt sure that here was an opportunity for this country to regain its leadership in this field. He produced a bit of paper with a squiggly line on it, and claimed that this was a record of radio waves which he had received with his first telescope, and which had been travelling through space for a million years. I was very impressed, and I was prepared for his view, which he pressed with considerable insistence, that what he really wanted was a huge telescope which he could point to any part of the sky and receive radio waves from any direction. He promised that if he had this, he would prepare a new map of the universe which would be even more complete than the one obtained by the great American astronomical optical telescope.

In the course of my experience, I have always doubled the amount which any enthusiastic scientific research worker says he wants for any new enterprise, so I said to myself that the amount required would be something like £200,000 or perhaps £250,000. I did not say anything then, but I asked my Scientific Grants Committee, on which we take care to see that there are some good engineers, to look into the proposal in all its aspects. The Committee's conclusion was that this telescope, consisting essentially of a huge suspended steel bowl, would weigh several hundred tons. The project was probably feasible but would cost a lot more than £100,000. After some argument, we gave this young man, whom you will have gathered is our lecturer, Professor Lovell (in spite of his disarming and innocent appearance), a few thousand pounds to spend on some good consulting engineers to go into the question of design in sufficient detail to determine the feasibility of the proposal and its probable cost.

So it was that the D.S.I.R. became aware of the idea of the telescope and at a meeting on 22 June 1950 agreed to give me a grant of £3,300 to enable Husband to detail and price the instrument in his sketch. This time we had no option but to approach D.S.I.R.—the University was still involved with the £1,000 under demand by the firm concerned in our previous efforts.

1950 was a year of intense activity. One short visit by Blackett to Husband's office in Sheffield was enough to satisfy him that I was now dealing with an engineer of high competence. Blackett himself suggested that the naval gunnery experts might help us with the problem of steering the telescope and the Admiralty's Gunnery Establishment in Teddington was one of the many places which I visited with Husband. This visit in time led Husband to the breaker's yard where he secured the 15-in. gun turret racks of the *Royal Sovereign* and *Revenge* battleships at a bargain price. They were dismantled and stowed in our workshop at Jodrell where they were to remain for many anxious years. The road between Husband's office in Sheffield and Jodrell Bank soon became a transit area for both of us and in his company I had glimpses of engineering processes and routines which soon made me realize that Husband had embarked on a mammoth venture. It soon became obvious, too, that he was doing the job because it was a great challenge and under financial arrangements which held little hope of profit for his company.

Husband had promised to draw up a memorandum with his outline proposals, general arrangement drawings, and price estimates by February 1951. There was a vast amount of work to be done before then both by Husband and ourselves. Part of his preliminary investigation was to find out what the foundation situation would be at Jodrell. This meant getting in a drilling rig—and there I began to bring a load of troubles on myself.

I should explain at this point that the few acres of land in the possession of the University at Jodrell Bank had already become inadequate for my purpose. We had 'borrowed' from the botany department the third field which had not been brought under cultivation, but I had already extended into a neighbouring farmer's field by renting a part of it from him on which we built one of our meteor aerials. Into a strip of this field we had sunk over fifty blocks of concrete on which to build our special aerials and we had erected a hut between them.[1] All went well until I tried to pay the farmer his rent. The vigilant eye of the University's finance department uncovered the irregularity. I was told very firmly indeed that land negotiations were the prerogative of the appropriate University officers and I was instructed to remove my aerials and hand the land back to the farmer. I doubt if there has ever been an instruction which I had less intention of carrying out. The building of the aerials had been a hard job and we were in the midst of the excitement of the daytime meteor streams. Alas! not for the last time had I acted impetuously and irresponsibly in order to get on quickly with the research and once again I owe my salvation to the Vice-Chancellor. One sunny afternoon Sir John said he wanted to come out to see what all the trouble was about. When I showed him in considerable apprehension what I had done and the results we were getting he said, 'Of course you can't move this—see if the farmer will sell us the land'. The final outcome was that the farmer agreed to sell the entire field and another one some distance away. This other field—still known as 'Field 80'—was the scene of my first troubles over the siting of the new telescope.

[1] These were the 'radiant' aerials with which the phenomenon of the daytime meteor showers were elucidated. The aerials and the blocks remained until 1964. The wooden hut was replaced by a permanent building and that hut formed part of the cricket pavilion at Chelford, until it was rebuilt in 1964.

Although the University owned Field 80 I had never regarded it as a good place for the telescope. For a variety of scientific and aesthetic reasons I wanted the telescope built in Field 132. This field belonged to a large neighbouring farm and inquiries had elicited that there was no hope of buying it. During the 1950 design stages of the telescope we had therefore to presume that it would be built in Field 80—the only site which we possessed which would take the instrument. Unfortunately we had only a footpath access from our main site to Field 80 and our only chance of getting Husband's drilling rigs there was to take them over land belonging to another neighbouring farmer. As soon as this situation became apparent I called on the owner of this farm, who was a pleasant middle-aged spinster running her small farm single-handed but depending for advice on Joe, a neighbouring farmer. Since the drilling was becoming urgent I thought that my chances of persuading Joe to give the right advice for transit of the machinery were reasonable.

Immediately the conversation began I received a severe setback. Joe didn't 'think much of them University people'. His antipathy arose because several of his cows had died some years before. Hopefully he asked some University people to test his land. I have no idea who came out from the University but they certainly created a bad image. Joe said that two 'young chaps' had casually looked at the land and that was the last he heard from them. In despair he had got other advice and an obvious source of lead poisoning was discovered. It took me a long time to build up a better image of ourselves and to persuade Joe to advise the lady farmer to let the drilling rigs over her land to Field 80. The delay was irksome and costly. Instead of going in over hard dry land the rigs made deep ruts in the land softened by the autumn rains and we had agreed to restore and pay for all damage. However, Husband's proposals for the foundations of the telescope were based on the results of these drillings in Field 80, although the telescope was not in the end built in that field at all.

8

Lobbying

I SUPPOSE that if I had been left to my own devices I would have given Husband's proposals to the D.S.I.R. in the spring of 1951. They would have been faced by a request for a quarter of a million pounds, whereas they might have expected a request for less than £100,000 based on my previous contacts with them. It seems quite certain that if this had happened the idea of the telescope would have died a quick death.

Blackett had much experience and wisdom in these matters and late in 1949 he began preparing the background which was to turn out to be so important in later years. He said that an essential condition was to obtain the backing of the key scientists in the country who might be interested or be consulted about the project. To do this he asked the Council of the Royal Astronomical Society to set up a committee—not primarily to back my telescope, but to 'consider what ought to be done to support radio astronomy in the United Kingdom'. At that moment there were three radio astronomy research teams active in the country: J. S. Hey's at the Royal Radar Establishment, Malvern, Martin Ryle's team at Cambridge, and my own. The three of us were asked by the Committee to table our proposals at its first meeting in London on 10 February 1950. I made a strong plea for the proposal to build a steerable paraboloid 250 ft. in diameter. Minute 11 of the meeting reads: 'Dr. Lovell then asked for consideration of his project for building in this country the 250 ft. parabolic receiver so mounted that it could be swung round to any region of the sky. Preliminary drawings and a memorandum were submitted. After some discussion it was agreed that this was a matter of great importance for the future of radio astronomy in Britain and that the matter should be placed on the Agenda of the next meeting.'

Since my proposal was the only one requiring special support the next meeting was quickly arranged for the specific discussion of this project. Sir Edward Appleton[1] was the Chairman of the Committee

[1] Sir Edward Appleton, G.B.E., K.C.B., F.R.S., Nobel Laureate, after a most distinguished scientific career, was Secretary of the D.S.I.R. from 1939 until 1949 and the Principal and Vice-Chancellor of the University of Edinburgh until his

and we agreed to meet in his rooms at the University of Edinburgh of which he was Principal. This was the critical meeting and I remember the anxiety as I drove away from my home late on a Sunday evening (26 February 1950) to get the night train from Manchester to Edinburgh. Fortunately Blackett and I met on the same train as arranged, and this ejected us at 6 a.m. on to a bitterly cold platform at Carstairs. In the cold, dawn journey to Edinburgh which followed, Blackett lost no time in going over our tactics at the meeting. For some reason he thought that my new memorandum was a 'disaster' but he 'hoped the Committee wouldn't think so'. Even more devastating, though, was his insistence that I should not put forward the proposal to the Committee on behalf of myself as something to be built at Jodrell, but as a project for the U.K. with the Committee to decide the best place to build it. All this and much more before breakfast on that slow, cold journey from Carstairs to Edinburgh! On many occasions in later life when I have reflected on these beginnings I have marvelled at Blackett's tactical mastery of the situation. I was like a wild animal, wanting nothing but the telescope, and I was to leave plenty of wreckage around me. Without Blackett's clearing of the ways all would have been wrecked at an early stage in the proceedings.

We had reached a critical moment in the negotiations for the telescope. I had no illusion that even a murmur of dissent from a single member of the Committee would have been fatal. I was asked by Appleton to outline my proposals, during the course of which I explained that the project was for a telescope for the U.K. and that it would be for the Committee to advise where it should be built and who should run it. It must have been a great effort for me to say that with any feelings of sincerity, but the Committee reacted like a gunshot and interrupted my discourse to say that they wished to make it clear that if any such device was to be built then a condition must be that it should be built at Jodrell Bank where the techniques existed and be run by the person who had proposed it. Blackett had instantly achieved his object!

After my explanation there were various questions, and then Appleton asked each member to state his final opinion on whether the Committee should back the proposal. I do not think I have ever had such moments of suspense, as one by one the members gave their favourable opinions. Finally Appleton said in words which I have never forgotten that he was 'impressed by the wide list of problems in astronomy and geophysics which Professor Lovell had listed as capable

---

death in 1965. My own early researches at Jodrell on meteors impinged on Appleton's historic ionospheric researches and he always exhibited the greatest kindness and helpfulness towards me.

of solution by a radio telescope of this size, but I am even more impressed by the possible uses of this instrument in fields of research which we cannot yet envisage'. In later years I have been given frequent causes for remembering these prophetic words. Without hesitation the Committee agreed to ask the Council to pass a resolution in the following terms:

> The Council of the Royal Astronomical Society strongly endorses the proposals put forward by the Physical Laboratories of the University of Manchester for the erection in the United Kingdom of a steerable paraboloid aerial of 250 feet diameter. The Council considers that by the erection of this apparatus the prestige of science in Britain would be considerably enhanced. In giving its support the Council places on record its opinion that the investigations to be undertaken are of high scientific importance including, as they do, the systematic survey of both the isolated centres and the general background of galactic radio emission, the study of the radio spectrum of the galactic and solar radiations, the extension of the meteor programme to meteors fainter than the 6th magnitude, the further investigation of auroral phenomena and the measurement of reflected pulses from the moon, planets, gegenschein and solar corpuscular streams. The Council is impressed by the consideration that the construction of the proposed paraboloid would permit the continuation in the United Kingdom of new methods of astronomical research, which have been so greatly developed by the skill of scientists in the United Kingdom, and which are independent of climatic conditions.

In this way Blackett obtained the support of the official body of the country's astronomers. With this backing and after further discussion with Husband, the University sent off my formal application to D.S.I.R. on 14 June 1950 for the sum of £3,300 for the design study. Eight days later, as mentioned in the previous chapter, the D.S.I.R. Committee gave authority for the grant to enable Husband to proceed with the detailed study.

In addition to the constant liaison and visits with Husband during the remainder of 1950, Blackett also insisted that I should draw up a detailed memorandum on the scientific aspects of the proposal to accompany Husband's engineering proposals. This took much time. It covered all aspects of the problem—the researches at Jodrell Bank and elsewhere which had led to the idea of the telescope, the tasks which the telescope could undertake, the scientific basis for the shape, size, focal length, and other data on which we had asked Husband to base his design, problems of siting, and power supply. The memorandum included everything which I could think of as relevant at that time and it was fortunate that J. A. Clegg was with me to write the chapter giving the calculations about the beam shape and power gain of the telescope. In addition J. G. Davies had been assigned the task of building a bench prototype of the analogue computer which was to supply the information necessary to drive the telescope in sidereal motion and he was responsible for the chapter on this matter. Eventually we typed and stencilled several copies of this memo which we bound in blue cloth so that it soon became known as the 'blue book'.

Blackett must also have carried out a large amount of background lobbying, the extent of which I have only been able to imagine in later years. At that period the only result of which I was aware was that every appropriate visitor to Blackett was despatched to Jodrell so that he could be suitably impressed by the new science which was developing so rapidly.

Finally, early in 1951, Husband's proposals and the blue book were ready. The way was now clear for a formal application to be made to the D.S.I.R. On the standard forms on which such applications are made the request for £259,000 to build the telescope was despatched on 20 March 1951. I suppose few occasions since the war have been more inappropriate for such a request to be made. The financial crisis in the country was causing the suspension of all but the essential and cheap D.S.I.R. researches and there was little hope of an immediate grant. Furthermore prices were increasing rapidly and some time after the application had been made there was a 20 per cent. increase in the cost of steel—implying at least a similar rise in the estimates for the telescope.

I have little doubt that at this stage the plans for the telescope would have been abandoned but for the efforts of the many influential friends who were working behind the scenes on its behalf. The full extent of what happened in that year of suspense has never been revealed to me, but it seems certain that Sir Henry Tizard,[1] Blackett, and the Vice-Chancellor never wavered in their determination to find some means of financing the project. Tizard, whom I had met frequently during the war, understood as well as anyone what effect a rejection of the project would have had on me, and I recall vividly an occasion when he was in Manchester to open new research laboratories for Simon Carves Ltd. Afterwards he asked to be taken out to Jodrell and during that journey he warned me in the kindest possible manner to be prepared for a failure since there were acute financial difficulties.

However, as the days of 1952 began to lengthen the efforts of these men and many others triumphed. The D.S.I.R. could not provide all

[1] Sir Henry Tizard, G.C.B., F.R.S. At that time Tizard was in the Ministry of Defence as Chairman of the Defence Research Policy Committee, and of the Advisory Council on Scientific Policy. An account of Tizard's distinguished career and of the vital role which the Tizard Committee played in the development of radar is given in *Biographical Memoirs of the Royal Society*, vol. 7, p. 313, 1961.

the money but the Nuffield Foundation had agreed to share the cost of the project, which was now estimated at £333,000. On a spring afternoon in March Blackett phoned and simply said 'You're through', and a few days later on Good Friday I received a formal letter from the D.S.I.R. announcing the award and the conditions attached to it. It was with joy and optimism, ill-matched to the troubles ahead, that I lived through those days.

9

Land troubles

THE proposals which had been accepted were based on a location of the telescope in Field 80 where the drillings had been carried out. However, as explained in Chapter 7, I could never visualize the telescope in this position. I wanted to build it in Field 132. During the year when the negotiations were proceeding for the finance of the telescope the elderly lady who owned the farm which included this field had died. She had several sons, all of whom had moved away to their own farms except the youngest who was living with his mother and farming this land. We learnt that this entire farm would have to be sold in order that the deceased lady's estate could be settled amongst her family.

Quickly I informed Rainford who was aware of my desire for Field 132, and with little hesitation he asked the agents to look into this question. Thereby began one of the most appalling series of troubles which I have ever faced, since I found myself enmeshed in an embarrassing family dispute. If I had left it to Rainford and to those who were capable of dealing with such affairs I would have saved myself much anxiety and I think the final outcome would have been the same. Alas, I was impatient. As soon as authority was given to build the telescope Husband began the process of obtaining estimates for the foundation work and obviously it was essential for him to know whether the telescope was to be built in Field 80 or 132; moreover, he was anxious to make a good start with the piling before the autumn arrived.

Since the actors in the saga are now living in happy relations with Jodrell it would be indelicate of me to relate the details of the private disputes which led to our troubles. It will be sufficient to say that Eric Massey,[1] who was resident on the farm and had been farming it for years, saw no reason why he should not continue to do so after the death of his mother. Formal and personal negotiations always ended in stalemate. But apart from the University's interest the legal position demanded a settlement and I spent many anxious hours visiting Eric and

[1] I am grateful to Mr. and Mrs. Eric Massey for their help in producing the story of this chapter in an acceptable form for publication.

the other members of the family. But anyone who knows anything about such affairs in local family communities will realize the endless tangles and frustrations which again and again hold up a near agreement.

Already, by May of 1952, the indecision about the site was causing serious concern. Husband had chosen the piling contractors and they wanted to start work. The formalities of obtaining steel and building licences had been completed and a starting date of 1 July had been officially allocated. Planning permission had still to be obtained and the documents could not even be dispatched to the local authorities until we had decided whether the site was to be Field 80 or 132. On 5 June the Area Planning Officer came to Jodrell Bank and yet another complication materialized. The boundary line between two rural district councils (and between two parliamentary divisions!) ran through Jodrell Bank. Field 80 was in one section, Field 132 in the other.

We were therefore exploring every conceivable avenue to break the deadlock over the purchase of Field 132, and one of my schemes was to swap 80 for 132. Rainford promised to come to Jodrell Bank with the agent to talk about this if I could persuade Eric Massey to attend. On Friday 13 June we assembled in my office, but the meeting developed in such a way that the agent (who was a tough individual) became so exhausted that he nearly collapsed and had to retire from the meeting. After this there was the utmost pessimism and on that day I wrote: 'It now seems to me to be extremely unlikely that we shall gain Field 132 in a year'.

But five days later Rainford's office phoned me to say that the Trustee's solicitor now thought that the transaction would go through and an answer was expected by the end of the week. Then on 23 June: 'The position could therefore hardly be worse and I said that it would be impossible to wait beyond the week; unless the negotiations reach a conclusion during the week they will be called off on Friday afternoon and we shall build in Field 80.' On Friday 27 June—the date agreed for the final decision—I could find no one in authority, but on Saturday morning 28 June Rainford said the position was hopeless and I phoned Husband's office to tell them to proceed in Field 80. Arrangements were made to begin the pile-driving in that field in a week's time.

On any reasonable assessment of the situation that should have been the decisive end but I was unhappy at a decision which I did not want. Within minutes of making the arrangement to proceed in Field 80 I was once more talking to Eric Massey and his wife, and they then agreed to one of the several proposals which Rainford had made on 13 June. After confirming with Rainford that he would still be willing to negotiate along those lines I returned to Eric and promised to visit all his brothers on the Sunday morning. I did so, but my efforts were in vain because by Monday morning Eric had changed his mind about accepting Rainford's offer.

So the tangle continued and on that day I once more toured the brothers and their solicitor. I was now moving in circles where I had no business and I still remember Rainford's justifiable annoyance and amazement when I phoned him from the solicitor's office on that day; but 'a series of miracles might still enable us to build in Field 132 this week, the first one being to assemble the brothers in Chelford at 2 o'clock'. Once more I toured the brothers to extract their promises to attend. The day was hot and I longed to join them in the hay fields where they were toiling.

The final manoeuvre to purchase the farm and the critical field then took a hopeful turn. Counsel's opinion was sought on both sides and as a result all the papers were prepared for a High Court decision on 24 July. But it was 1 July, and precisely on time the piling contractors arrived to begin work. They started boring in Field 80.

On the day of decision—24 July—I had to lecture in Leeds and the issue of the field was so vital that I had arranged for the solicitor to communicate with me there. At 4 p.m. his telegram simply said 'Leave to sell granted'. I phoned Husband's office and said the piling operations must be transferred to Field 132. Alas! the High Court settled the legal problem but not the personal one. We were far from the end of the road—signatures were required and legal rights notwithstanding families tend to cohere when final disruption is threatened. On 19 August the solicitor phoned to say that the eldest brother had asked to see me again. After much talk he added his signature to the contract. '*Wednesday August 20*. I came away with the contract in my pocket hardly believing that this would be the end of the affair.'

It wasn't! That afternoon more difficulties appeared: who was to carry out the act of possession—the brothers or ourselves? (Eric had refused to be a tenant.) 'There seemed no alternative but for me to return for another interview with the eldest brother. This time I found him on top of a stack with a load of oats beneath him.' After the cartload was deposited on top of the stack we walked towards the house and got the solicitor on the telephone. 'I have no emotions left—the whole matter has dragged on so endlessly that I could not believe that other factors would not arise.'

> *21st August*. [The solicitor] replied like a new man—he said we could now take possession. . . . this really does at last look like the moment when the first stage of acquiring the farm is at an end and that tomorrow we should be able to enter the field with confidence.

22nd August 1952. Just after 2 o'clock the long awaited assembly occurred for the marking out of the access road and the perimeter of the radio telescope. . . . In Field 132 the site of the telescope was located as far as possible in the north westerly corner leaving about 20 ft from the fences. The control building was sited in the SE corner facing towards the paraboloid. . . . The positioning was done without any interference from Eric. . . . It was indeed difficult to believe that this sunny afternoon marked the real beginning of the work on the radio telescope with all our requirements over site and access fully met.

The belief was illusory! At 2 a.m. on the morning of 26 August one of the night staff at Jodrell phoned to say that a tractor was operating in Field 132, and that the contractors' hut had been demolished and thrown into the ditch together with all our marking-out pegs. Moreover, daylight revealed the presence of a ferocious bull in the field. I summoned the police. 'We walked to the scene of the ditching of the hut and were immediately approached by the bull causing us to retreat. The police were incredulous and somewhat uncertain what action to take.' They returned to their H.Q. with a promise that they would give me their advice by phone. To my amazement the police said they were unable to do anything because it was entirely a civil matter. '*August 28* . . . the situation over Field 132 remains the same, namely the bull in possession and the hut in the ditch.'

In retrospect it is easy to understand Eric Massey's fears and anger, which were probably increased by our impatience, but the University, when the legal formalities were completed, approved all the offers made by Rainford to Eric Massey and it must be said that since that date it could not have had a better or more co-operative tenant. To us he has been a very pleasant neighbour and the early arguments seemed to have cemented extremely friendly relations between a good farmer and a research station.

10

The design of the telescope

AT this stage it is necessary to say something about the telescope as it was designed to be built at that time. The blue book put forward many reasons why it should be at least as large as the transit telescope and we eventually decided to ask for a paraboloid of 250 ft. diameter, 32 ft. larger than the aperture of the transit telescope. There were two other vital design factors of the paraboloid which had to be settled: the focal length and the accuracy of the construction. The transit telescope was of long focal length; as already explained, the bowl, for purely practical reasons, was shallow and the focus was at the top of a 126-ft. steel tube. This arrangement exposed the primary feed aerial to unwanted sources of interference and also seemed likely to present unnecessary problems to the engineers. Therefore we decided to increase the depth of the bowl and finally we asked for a paraboloid with the focus in the plane of the aperture. For a 250-ft. diameter bowl this meant that the engineers would have to provide a 62½-ft. tower, rising from the apex of the paraboloid, on which we could mount our aerial systems.

The accuracy to which the paraboloid was to be constructed and maintained at all operating angles was of fundamental importance to the engineers; since on this would depend the stiffness of the instrument and hence the weight of steel and cost. During the years when we were discussing these problems the issue appeared to be quite straightforward. The radio emissions from space had been studied only on relatively long wavelengths and the only known sky surveys were those of Reber made on a wavelength of 1½ m. with a beamwidth of 8°, and of Hey, made on a wavelength of 4 m. with a beamwidth of 13°. Our own transit telescope had surveyed the zenithal strip at much greater definition on 2 metres where its beamwidth was ± 1°. There was negligible experience on the reception of the cosmic radio waves on wavelengths shorter than 1 m. It was therefore a commonsense decision to take a wavelength of 1 m. as the shortest working wavelength of the telescope at which its beamwidth would be ± ½°. With a working wavelength range of 1 to 10 m. the telescope would produce sky surveys of far greater definition than any other one available and would be a pre-eminent instrument in all our other activities in that waveband.

This decision immediately led to the specification that the shape of the paraboloid bowl was to be within ± 5 in. of the true shape relative to the focus. Further, the size of the mesh of the reflecting surface was determined by this shorter wavelength limit and we specified that it should be formed with 2-in. square mesh. With these parameters the efficiency of the paraboloid was expected to approach the theoretical value closely at the shortest operating wavelength.

The basic problem which we presented to Husband was therefore to build such a giant paraboloidal wire mesh bowl and mount it so that we could drive it separately in elevation and azimuth to give full sky coverage with continuously variable speeds from 2 degrees per hour to 36 degrees per minute with an accuracy of $\pm \frac{1}{5}°$. We also specified that the telescope could move in sidereal motion, that is simultaneously in azimuth and elevation with the rates determined by a computer so that the instrument would track a star or planet automatically. Our detailed specification was loaded with a mass of miscellaneous requests concerning the detailed nature of the controls, cabling, and access problems, etc.

All of these requirements were contained in Husband's report of February 1951 in which he proposed to meet our demands by driving the bowl in elevation through the two battleship gun racks mounted on trunnion bearings supported on two steel towers 186 ft. high. The azimuth motion was to be obtained by mounting the towers on 12 bogies, 6 under each tower, running on a 320-ft. diameter double railway track. One acute difficulty which we faced was how to obtain access to the focal point in order to change our aerials. We asked for the bowl to be capable of complete inversion so that the aerial tower was pointing to the ground. Then we asked for a mobile electrically driven tower to be provided on the top of the cross girder, normally stowed at one end but capable of motion along the girder to give us access to the aerial on inversion. This tower and many other features of Husband's original design are clearly shown in illustration 6.

The cost of this instrument was at that time estimated as just over £200,000. To this was added another £30,000 for an extension to the power house, provision of the control building, and miscellaneous roadworks. At the end of his report Husband wrote, 'We anticipate that the whole scheme could be in operation within two and a half years of work commencing. This estimate of time takes into account the delivery periods quoted by the various contractors who have submitted tenders.'

There was already alarm at the rapid increase in the money which I was requesting. In 1949 I was talking about a figure of £50,000–£60,000. At the time of the Edinburgh meeting in February 1950 I find that I estimated the cost as between £50,000 and £100,000. The first formal application to the D.S.I.R. on the basis of some real estimates by the engineers was for £154,300, the next application on 20 March 1951 after the more detailed investigation of the design was for £259,000. In the succeeding year the general rise in prices increased this figure to £335,000 without a nut being changed by ourselves or the engineers. How little did I realize when this grant was awarded that we were merely on the fringe of our financial troubles.

11

The foundations of the telescope

On 3 September 1952 lorry-loads of sand and gravel began to move into Field 132. The tender of Messrs. Whittal of Birmingham for the foundations with the Cementation Company as sub-contractors for the piles had been officially accepted. The episode of the bull and the demolition of the contractors' first hut had left a suspicious air about the place and the contractors refused to start work until they could be satisfied that they had a legal right to do so. This was soon resolved by sending a messenger to the University to bring out as quickly as possible an official letter from the Bursar's department to Whittals assuring them that the University had full legal possession of the field.

During the next few days the first pile was sunk near the region of the central pivot. There was 40 ft. of running sand, then a mixture of clay and sand, and at 55 to 60 ft. wet red marl. One hundred and sixty similar piles were required to support the central foundation block and the circular railway track. The procedure was to erect a rig over the pile position and drive in successive lengths of steel casing about 18 in. in diameter. The sand and water was extracted by a bucket which was lowered into the tubing at short intervals. When the bottom of the steel casing reached the marl, reinforcing rods were inserted, concrete was poured in and finally the steel casing withdrawn. After experience with driving the casing for the first pile it was agreed at a meeting on 16 September that Cementation should complete the 160 piles in ten weeks so that Whittals had a reasonable chance of completing the foundations before Christmas. It was not the only target date to go awry.

September 18. 5 rigs for pile driving arrived and being erected today. The first bore hole has still not been concreted neither has driving of any of the others started.

September 20 (Saturday). 3 rigs working and the foreman said that on Monday they would be working with 6 and with good weather the piling should be finished in 10 weeks—for some reason which I do not understand the men work until 4 p.m. on Saturday afternoons. At 3 p.m. there were 20 men on site, 10 in working clothes, the others dressed to go out for the day.

Of the 10 in working clothes, 4 were actually working on a pile which had reached 50 ft. Water was pouring into the bottom but the foreman said they were driving in another length of casing to seal the bottom and thus save a fresh start on Monday. At 3.30 the casing was screwed on but after three small taps a man appeared with instructions to stop work. They downed tools (at 3.40) and left the water flooding into the hole.

It was the first of many occasions when I was to find myself alone on a deserted site wondering at the ways of British workmen.

September 23—the trouble at the moment is that so much water accumulates in the holes overnight that it takes them 3 hours the next morning to clear it out . . . this is losing 3 hours per rig per day.

The piles were going many feet deeper than expected and by the end of the month only eight were finished and all had gone deeper than 50 ft. before entering the marl. Now the hope was that the piles would be finished by the end of the year and the whole foundation by the spring. Activity on the site began to build up rapidly during October. The completion of the six piles under the central part of the telescope enabled Whittals to proceed with the excavations for the thrust block and cable trench leading to the control room. By 10 October twenty-six piles had been completed.

. . . as the pile operations move round the perimeter of the paraboloid it has become clear that the marl is extremely steeply shelving. The most recent pile in the SE struck red marl at only 46 ft whereas one in the NE went down to 60 ft without reaching any marl . . .

21 October—Piles 38, 39, 40 being put in today—the central thrust block is now being worked on steadily . . . for the first time it is possible to see the size of the annular chamber and the central concrete block marked out . . .

31 October—2 piles completed making the total 57. It was hoped to complete a third pile to the north but the marl wasn't reached until late in the afternoon at 68 ft . . . now a considerable worry since it appears that in the northern sector of the perimeter the marl is at about this level and presumably we shall be involved in a great deal of extra expense.

Progress became more and more difficult. Broken land drains added to the water troubles, the earth walls of the thrust block cavity collapsed before the concrete could be got in position, and in mid November I wrote that 'the weather remains wet and windy and the whole site is a quagmire'. The hundredth pile was finished on 3 December, and a giant excavator was carving out the earth from the top of the piles so that the shape of the ring beam was becoming visible. Unfortunately the piles were being driven deeper and deeper in the northern sector and early in December one had gone to 83 ft. before reaching the marl.

December 10—the main trouble is the great depth to which these piles are now being sunk. In the depression, marl is not being encountered until over 80 ft and the piles on the banks of the depression are having to be sunk to 75 ft . . . a large amount of reinforcing steel is in position along 50 yards of the tunnel and about 100 yards of the ring beam is in process of excavation . . . *December 12*—today's pile is right in the centre of the depression and has gone down to 89 ft. It now seems clear that this depression must arise from some very deep subterranean fault . . . the [second series of load] testing of the pile has been delayed because of the fiasco over the loading arrangements. [The 100-ton blocks of concrete used for the previous tests were no longer obtainable and it was therefore agreed to procure 110 tons of steel from S. Wales.] . . . unfortunately these were despatched in 9 railway trucks addressed to Jodrell Bank, Holmes Chapel, Nr. Doncaster. So far only 7 of the trucks have been located and these are in a small siding near Doncaster.

At the end of 1952 the score of completed piles stood at 108 and the hopes of finishing the foundation work even by the spring had vanished. In truth there had been many occasions since September when I was thankful that we had made even that amount of progress because the rumblings of the legal dispute over the farm and its tenancy continued.

During January the pile and other foundation work continued slowly under bad conditions.

. . . the site is full of incredibly sticky mud . . . [Jan. 14] . . . during the last 9 weeks only 9 piles have been completed . . . [Jan. 23]. *Jan. 28*—with Husband on the site—it is now possible to walk down the steps of the thrust block and although there is a good deal of water on the floor and it is still full of supporting rods, one can walk inside the annular chamber . . . in the north east region the piles are penetrating 90 ft before meeting the marl . . . Husband not at all worried about the slowness in finishing the piling because these operations are fairly far ahead of any progress which can conceivably be made by Whittals in the next few weeks . . .

After some February blizzards the site dried up and early in March we were finding it easier to count the piles which still had to be done rather than those completed. '134th pile just finished, so that there are only 18 now to be done . . . on the ring beam Whittals are now excavating for the fifth section, they will therefore soon be half way round. [March 5.]' But the progress was agonizingly slow and far behind any target date ever set.

. . . Husband came at lunch time, mainly to inspect the work on the foundations, he is very unhappy with the progress [March 19] . . . Only one man is now working and he is scheduled to do two per week and does not appear to be hurrying himself to do more [March 21] . . . the ground on the site is now caked hard and the foreman is beginning to grumble about the shortage of bricks and cement [March 25] . . .

April 2, 1952 (Maundy Thursday) . . . quota of 2 piles completed making 142 and Whittals have finished the ring beam between the first lot of expansion joints . . . it has taken them 2 weeks to do the concreting in one section and even at that rate of progress it will be 6 months before the ring beam is finished . . . Last Easter was the time when the announcement of the grant for the radio telescope was made, a year later on this stormy afternoon, the site was almost deserted. It is piled with mounds of earth, and on the whole is somewhat depressing. There is little to show for the vast amount of steel and concrete which has already been sunk into the ground . . .

Cementation are on the very last pile. Unfortunately this is the one in which they have lost the pumping tool [May 12] . . . Cementation are still fishing for the bucket, which is 50 ft down in the last pile [May 15] . . . managed to recover the bucket from the last pile yesterday afternoon [May 19].

On 21 May we heard the last of the pile-driving hammers. An operation scheduled in September to take two and a half months had lasted for nine months—a ratio of promise to achievement with which we were to become wholly familiar. But the completion of the piles did not mark the completion of the foundations. By mid June Cementation had gone from the site but Whittals had still only completed half of the ring beam. In fact it was 31 July before the last gap in the 350-ft. diameter circle of concrete of the ring beam was closed, and Christmas Eve before Whittals finished the tunnel from the central pivot to the site of the control room. Ten thousand tons of reinforced concrete and nearly £50,000 had gone into the ground. After sixteen months' work, all that was visible was a 20-ft. wide, 350-ft. circle of concrete.

The slow progress on the foundations concerned me less and less as the year progressed—I became submerged in a multitude of other anxieties and difficulties which threatened the whole concept of the telescope. The question became not whether it would be finished by 1954 or 1955, but whether it would ever be built at all.

12

The first financial crisis

THE increases in the estimated cost of the telescope during the design and negotiating stages have already been described. In the spring of 1952 when the official announcement of the grant was made the cost was estimated to be £335,000 and nobody was prepared for, or expected, more than a 10 per cent. deviation from this figure. Personally I had no competence or experience to handle financial problems. In the enthusiasm of that Easter period I felt that I had already had more than enough of the finance of the telescope and that only the excitements of construction and use lay ahead. I have often pondered subsequently what I would have done at that time if I had known even a little of how utterly wrong that belief was, and that nearly ten years of financial nightmare lay ahead during several of which every dawn carried with it the possibility that the project would be abandoned.

The detailed financial procedures were agreed at a meeting between the D.S.I.R., Husband, and the University a few days after the notice of award of the grant. Rainford was to place the actual contracts on receipt of agreed specifications between Husband and ourselves. The payments of the accounts were to be settled directly between Rainford and the D.S.I.R. It was also agreed that for contracts of £5,000 and over the prior consent of D.S.I.R. would be required. In effect this meant that almost every item for the telescope had to be referred to the D.S.I.R. before Rainford could take formal contractual action. At this meeting it was agreed that competitive tenders must be obtained wherever possible, but on the advice of Husband it was agreed to proceed with the United Steel Structural Company Ltd. alone for the main steel superstructure of the telescope.

In the beginning everything went smoothly, the lowest of the various tenders were accepted for the foundations, the power house extension, and the control building. The first glimpse of impending difficulty which I can find in my diary is on 23 October 1952 where I record a visit to Jodrell by Rainford who seemed 'quite happy about the state of things. His main worry was the lack of the contract for steel'. That afternoon I drove over to Sheffield to see Husband about many things and on that subject he said that he was still waiting for the approval of the draft contract from the United Steel Structural Co. Ltd. On 2 December I record that Rainford was still worrying about the contract and that I promised to make more inquiries during a forthcoming visit to the steel company. However, my notes of that visit on 10 December are full of pleasure at the plans for the fabrication of the steel and delivery and erection at Jodrell to meet the target for completion in 1955 but contain no mention of any discussions about the contract.

In mid January I find a reference to the fact that Rainford was now not merely worried, but considerably annoyed about the steel contract and I don't think that my conveyance of these sentiments to Husband helped. A few days later Husband wrote to Rainford about this to say that there had been much difficulty about the special steel for the railway track and it was impossible for them to be hurried on a project where so much was at stake. In any case the appropriate documents were sent to the Bursar soon after that and on 23 January I mention a phone conversation with the Bursar's assistant 'worrying about certain aspects and also the increase in price . . . estimates that the total amount specified in the document would come to £109,000 instead of the £105,000 allowed and there were several items expressly excluded from the contract, which do not appear to be included elsewhere'. This seemed to me a minor matter of a few per cent. well within our safeguard of an estimate to 10 per cent. On 30 January I wrote a long letter to the Bursar giving my point of view on the increases in cost. The original figure of 959 tons of steel was based on the provisional drawings of February 1951. Now the structure had been detailed the tonnage had increased to 1,000. I agreed with his view that the cost of the steelwork was now £120,000 instead of £105,000, but I argued that the figure was still in line with the original estimate, since a 17½ per cent. rise in steel prices had taken place. However, I promised to explain the situation to D.S.I.R. and arranged a special visit to do so.

February 20 1953. Husband came over yesterday . . . and the day was taken up with a detailed price review so that I could present the picture to the D.S.I.R. on Tuesday. The overall result was absolutely shattering . . . the net result being that on the capital expenditure we require a 32% increase of over £92,000 . . . 20% of this is due to rises in engineering costs since the original schedule and the remainder is due mainly to the fact that the weight of the structure has now come out at 1200 tons instead of 1000 tons in the original plan.

On 25 February I faced the D.S.I.R. officials. At this meeting it seems that my line was to argue that the main increase was due to rising costs and that we had in fact 'done quite well to keep the other increases within about 10 per cent.'. Since we had underestimated the cost of the mesh by 67 per cent. and of the central pivot by 100 per cent. the basis of my argument seems a bit thin at this distance in time. Anyhow D.S.I.R. did not have that amount of money to dispense and it was obvious that another approach to the Treasury was inadvisable. 'Evans[1] promised to do whatever he could, but saw little hope of obtaining an answer under two or three months. We thus enter a further phase of acute anxiety as to whether the telescope can be constructed at all in its present design.'

If there was to be any progress in getting this additional money we clearly had to provide D.S.I.R. with a more detailed story of why we needed it. The truth was, of course, that no one had ever attempted a similar structure before and numerous additional matters were occurring as the design became more and more detailed and as estimates turned into actual tenders. By mid March Husband had sent us an analysis of the main increases in tonnage. For example, after more detailed analysis of the stresses he had strengthened the ends of the back girder of the bowl by 30 tons each to give more torsional rigidity. Alas, 30 tons in the cross girder and 10 tons on each tower were for extra strengthening on account of the extra windage arising from my request for a modification to the central part of the mesh (see page 83). Various other minor increases had raised the weight to 1,177 tons, 562 in the bowl and 582 in the supporting structure and 33 tons auxiliary apparatus. At this stage Husband was able to say that the United Steel Structural Co. Ltd. had checked the tonnage and agreed closely with these figures from which there was unlikely to be any substantial variation.

In discussion of the overall time scale Husband still holds to the opinion that the job will be largely complete by the end of 1954. Hence, it still seems reasonable to anticipate that the instrument will be working by about the middle of 1955 [19 March 1953] . . .

March 31st [referring to a letter from D.S.I.R.] . . . asks us to complete an application form for the Grants Committee . . . the most depressing part of the letter was a statement that the Nuffield Foundation felt unable to increase their allocation by more than another £32,000—making £200,000 and this meant that the D.S.I.R. would have to find £60,000 . . . This was going to be an extremely difficult matter, and it would be a great help if the University could demonstrate its backing of the project by making a contribution.

These difficulties were now causing serious delays because Husband was unable to progress the steel work until contracts were signed. On

[1] I. G. Evans, C.B.E., Assistant Secretary in D.S.I.R. Headquarters. In charge of grants until he retired in 1953. He was succeeded by Cawley (see p. 77).

the financial side Rainford and the D.S.I.R. refused to make another move until they could be quite certain of the new overall contract figures and it was 23 April before contractors' tenders for all the major items were available (the last to come being that for the control building).

The total revised highest estimated expenditure was £459,896—a 33 per cent. increase over the money already allocated. In a covering letter to D.S.I.R. explaining the reasons for this increase we said that the breakdown of the increase was 10 per cent. estimating errors, 14 per cent. increased costs, 2 per cent. items not originally included, 2·4 per cent. painting and resident engineer, 3·9 per cent. increased professional fees, and 0·7 per cent. increased salaries. Armed with these final figures Rainford met the D.S.I.R. finance officials on 28 April. He found them 'helpful and anxious to overcome the present dilemma', but the practical result was that we had to make a formal application for the meeting of the Scientific Grants Committee of D.S.I.R. on 6 May.

Although Rainford thought our case was reasonable and that there might only be a few weeks' delay in obtaining a further authority to proceed, nevertheless these were gloomy days. Husband was disturbed about the delay in placing the steel contract. On 4 May he told me that the situation was critical. The tender submitted by the United Steel Structural Co. Ltd. was 'no different from what it would be if the same tonnage of steel had to go into building a bus station, and that if there was much more argument about the contract it was conceivable that they might withdraw their offer to undertake the work'. Blackett, too, was intensely annoyed with me and said that if there was too much delay in completing the telescope there was a very grave danger that the cream of the programme might be skimmed in other parts of the world with less ambitious apparatus.

Then the meeting of the D.S.I.R. Scientific Grants Committee on 6 May to which Rainford and I were called went awry right from the start:

To begin with the Committee was quite different from the one which originally recommended the award of the grant. Sir George Thomson[1] immediately seized on the wind loading problem as it affected the extra steel and took us back four years to arguments about the wind tunnel tests . . . finally it was agreed that we should obtain an assessment of the position from an independent aerodynamic expert—Prof. W. A. Mair was suggested as one who had earlier been consulted about the project. [The meeting emphasized that the structural calculations of the consultants were not in

[1] Sir George Thomson, F.R.S., Nobel Laureate, was at that time Master of Corpus Christi College, Cambridge. He had retired from the Chair of Physics at Imperial College (to which Blackett succeeded) in 1952.

question—only the basic wind pressure assumptions.] . . . The gloom was further deepened by an emphatic appeal that the University should show its backing by making a financial gesture. A violent and stubborn argument ensued between Rainford and the Committee . . .

(Rainford's case was straightforward; the University had no capital with which to help the project and that it was already heavily committed to Jodrell by way of staff and running expenses, etc.)

. . . I have rarely attended a meeting which took a direction so different from that which I expected, or one which had such a depressing outcome . . . it seemed to me that we had spent nearly £100,000 and yet were back in 1949 and 1950.

Later in the day I returned to D.S.I.R. and derived comfort from the permanent officials. Evans said that in spite of the attack made on me by the Committee he thought they were basically of the opinion that the project should proceed as quickly as possible. He tried to phone Mair for me but it was the evening before I could converse with him. In the early days of our thoughts about the telescope Mair was the director of the fluid motion laboratory in Manchester and I had often discussed the problem with him and he had made some calculations for us. Subsequently he had gone to a chair in Cambridge and now I was to reap the benefit of this early friendly association with him. As soon as I told him our troubles he wondered what all the fuss was about and was quite prepared to write a report on the basis of his original calculations.

Fortunately Husband was in London and the next day over lunch I carried out the unpleasant task of explaining this new development. Understandably he was greatly displeased but nevertheless for the sake of the project he offered to return immediately to Sheffield and draw up a report on the wind loading for Mair's consideration. Mair agreed to all of this but then, once more at D.S.I.R., I found that Evans was urgently in need of the document in order to prepare it for the meeting of the Advisory Council on the 20th, so that a new and immediate approach could be made to the Treasury. Because of this I phoned Husband who had now returned to Sheffield and asked him to send the report directly to Mair.

When I eventually returned home from London late on Friday night there was an urgent call from Husband who on thinking about the wind problems had decided that the only feasible way of handling this was to draw up a document and hold a personal discussion with Mair. Sunday in Cambridge was the only hope for a week and fortunately Mair agreed to see us on Sunday afternoon.

Sunday May 10. Met Husband in Market Harborough for lunch and discussed the document. Proceeded to Cambridge for the afternoon discussion with Mair, who readily agreed to the statements made. Mair considered that the design factors were perfectly reasonable and signed the document which I then took through to London.

Monday May 11. 10 a.m. handed the document to Evans at D.S.I.R. who thought it would meet the difficulty perfectly . . . seemed quietly confident that a favourable outcome might result by the end of the month.

The next day I wrote to Sir George Thomson:

Jodrell Bank
May 12, 1953

Dear Sir George,

Following your suggestion at the meeting of the Scientific Grants Committee on May 6th, we have prepared a short note regarding the wind pressure on the radio telescope in consultation with Professor Mair. I have handed the original, signed by him and the Consulting Engineer, to the D.S.I.R. No doubt you will be receiving a copy officially, but I should like to take this opportunity of sending a copy to you with some additional observations.

You may feel after reading the note that we have erred on the side of too large a safety factor. A wind speed of 134 m.p.h. is required to overcome the dead weight and anchorage, with the bowl in the worst position. This gives a factor of safety of two on the basis of the average maximum wind specified for this district in the B.S.S. document, and a factor of 1·6 on the abnormal wind conditions experienced here last winter. We can achieve a further factor of two by the anchorage arrangement, but we obviously cannot rely wholly on this—for example we might get a power failure at the critical moment. Apart from this the strength of the structure is also determined by our specifications regarding distortion of the bowl at very much lower wind speeds. We have previously considered this in great detail and unless we are prepared to have the instrument unusable on certain frequencies on moderate wind days we cannot reduce the strength.

Mair is perfectly satisfied that the basis of the wind pressure used in the calculations is correct. In this respect two points are particularly important. (i) In 1949 we discussed the question in detail with Mair, who calculated the forces on the mesh from basic data. The subsequent N.P.L. tests on the sample mesh confirmed his calculations in a very satisfactory manner. (ii) Only about 50% of the wind loading arises from the mesh itself. The remainder is on the steel structure which is of the type for which information is well established.

As you know, several years of calculation have gone into the specification and design, and the note attached is only a very brief summary of the salient features of wind pressure. We have planned another memorandum in which this work will be described in detail, but it will naturally take some considerable time to complete. I very much hope, therefore, that the present independent assessment by Mair will alleviate your concern.

Yours sincerely,
A. C. B. Lovell

A favourable outcome was absolutely vital at this stage and it was with great relief that I received Sir George's reply on the following Saturday:

Dear Lovell,

Thank you for your letter and for the report. I am entirely satisfied. I did in fact ring up Mair and had a short talk. He assures me that the maximum stresses will occur when the telescope is horizontal and that in these conditions the chances of aero-foil action are negligible. I am sorry to have given you the extra trouble, but I think for your own sake it is as well to have the report signed by an aerodynamics expert just in case anything happened.

Yours sincerely,
G. P. Thomson

The critical meeting of the D.S.I.R.'s Advisory Council was on 20 May and the next day Evans phoned with good news that the Council had agreed to provide us with the extra money. Alas there could be no joy because:

Evans said that it must be known that the Advisory Council was concerned about the large increase and gave the warning that it may not be possible for the D.S.I.R. to provide the necessary money for any further such increases in cost should they arise. In the unfortunate event of more being required it may be necessary for the University to raise it by appealing to another source.

Rainford's reaction was even fiercer than I feared: '. . . the job cannot possibly go on under these conditions'.

While we were floundering around in these financial arguments Husband was pressing the United Steel Structural Co. Ltd. On 19 May he had agreed with them that 'provided the contract could be placed within the next few weeks—delivery of steel would commence in February 1954 and be complete by May . . . gabbards and cranes could be erected in January 1954; the steel work erection would begin in March 1954 and be completed in September 1954.'

But how to get permission to place the contract? I thought up schemes for removing objections—for example, that we would delay the control building and thereby keep £30,000 in reserve: '. . . my attitude was that a control building without any steel work was useless, but that the steel work without a control building would still be a great deal of use.' This helped to the extent that in early June Rainford was willing to place the steel contract on this basis.

At last I thought all must be clear; the contract could be placed. Alas, the final details of the contract could not be agreed. 'We are now in the paradoxical situation of having been given permission to spend the extra money, and yet still not having a steel contract which can be agreed.'

Worse followed. By mid June it began to emerge that the United Steel Structural Co. Ltd.'s price was now even more than that for which we had just got the extra money—£5,805 in fact. On 24 June we all met in Husband's Sheffield office and it seemed from this inquiry that most of the new increases were justified in terms of increased labour costs, etc. By 6 July Husband had successfully negotiated the removal of some of the increases with United Steel Structural Co. Ltd. and the excess came down to £4,000.

At last on 18 July these protracted negotiations came to an end and we received authority to go ahead with the contract. Throughout the whole of this the D.S.I.R. officials showed much personal sympathy for me. 'Although there are some minor matters to be settled we do not wish to hold up the placing of the contract as we have no doubt that Lovell must be very worried.' Indeed I was beginning to need all the moral support I could get and it was fortunate that the storm clouds of the future were still below the horizon. At least by the end of July the steel for the telescope was being rolled, all partners to the contract seemed satisfied, and Husband 'said that everything was in order about the steel and that there was every intention of carrying out the erection next year'.

13

Railway tracks and power

In the spring of 1953 when we were preoccupied with the increase in the tonnage and cost of the main steelwork, there were a multitude of other less costly but urgent items to be dealt with. In March Husband was still negotiating for tenders on the control building, the railway tracks, the bogies and gearing, the main bearings, the driving and control systems, and the mesh.

The railway track on which the entire telescope was to rotate was clearly an item of great urgency since until it was in place on the concrete foundations no erection of any kind would be possible. The requirements were stringent. The track had to be level over the entire circle to one eighth of an inch. The diameter of the inner rail circle was to be 319 ft. and of the outer 353 ft., and double rails had to be used because the largest standard gauge was not strong enough to take the weight of the telescope. The outer of the rails had to be $\frac{3}{64}$ in. higher than the inner. Expansion joints were, of course, essential and these had to be disposed so that not more than one wheel of the 48 bogey wheels passed over one of the joints simultaneously. Husband originally intended to use specially hardened steel rails but the tender for this of £19,000 was far outside the estimate and it was fortunate that T. W. Ward were able to tender for a rail system in ordinary steel at an acceptable figure of just under £11,000. Before the end of June we had authority from the D.S.I.R. to proceed with this contract, and a few days before the ring beam was finished on 31 July Rainford was able to tell me that the official instructions to Husband to proceed with T. W. Ward had been sent.

It was fortunate that the placing of this contract was not seriously delayed by the current controversy over the major financial problem of the steelwork because Ward's had already indicated that they could not start work for three or four months after the receipt of the contract.

9 Dec 1953 Husband said that the tracks had been successfully drawn and were now in process of being bent to the correct curvature. . . .

11 January 1954 . . . Some of the sleepers for the railway track arrived on

Saturday and today the first ten rails have been unloaded. Wards have already bolted down about twenty of the sleepers. *21 January*—about three quarters of the inner circle is complete

but on '29 January . . . Wards are encountering big difficulties in laying the railway track because all the holding-down bolts are frozen. They were using a flamegun in an endeavour to melt the ice . . .', and on 1 February '. . . the holding-down bolts are frozen 2 ft into the concrete . . .' (at this stage the bolts were loosely held in holes in the concrete ring beam and these holes were than packed with ice). 'February 17 . . . all the rails are now in position except the closing ones over the tunnel and these are to wait until the exact lengths are determined by measurement.'

On 15 March I recorded that Wards had completely laid, surveyed, and adjusted the track and that concreting of the bolts would begin in a few days. The grouting-in process of the bolts continued steadily and without incident in the spring weather and in the event the railway track was complete long before anything was ready to erect on it or over it.

Another item had been moving forward steadily during this time, largely free from the troubles which assailed the steelwork. The first permanent buildings at Jodrell Bank mentioned in Chapter 4 included a power house for the diesel generators and by this time the battery of small mobile generators had been replaced by two 105 kW surplus generators. To provide power for the telescope it was necessary to extend these supplies and in the estimates we allowed for an extension to the existing power house in order to house a new 240 kW diesel generator. In July 1952 we went to Mirrlees, Bickerton and Day near Stockport where we saw in operation a diesel generator originally intended for delivery to Spain. The order was cancelled and since the generator with its automatic controls was precisely what we required at Jodrell Bank we settled for it immediately, thereby presenting ourselves with the problem of storing it until the power house extension was available—and this was at that time still on the drawing board.

By early September 1952, when we were deep in our troubles over Field 132, four contractors had sent in their tenders for this building work and since Whittal's, who were already designated as the contractors for the foundations of the telescope, were the lowest, their tender was accepted. There was a large number of minor rumblings over the costs mainly because I wanted to include a number of miscellaneous but associated items under the heading of the 'power house estimate'.

22 October 1952 . . . hold up on the foundations to power house because of difficulty in getting the 2000 buff facing bricks . . . cannot get any promise of delivery until mid-December. . . . phoned Butterworth [the University Buildings Officer] and pointed out that these were the same bricks as those used in the chimney under construction for the University heating chamber.

The bricks arrived and the power house extension proceeded. '*March 14*—the walls of the power house are nearly to the roof.' As for the generator problem we must have persuaded Mirrlees to keep it for us because: '*April 23* . . . the roof of the power house is being put in today. The generator has been booked for delivery on June 1st and is to be got in before completion of the floor.' The large cable from the generator to the telescope site had also been agreed and on that day '. . . they are measuring for the main power cable; and expect delivery five weeks after placing the order, so there seems to be little doubt that we should be able to get this in during the summer.' Early in August the generator was having its preliminary tests, the main switchboards arrived in October, and although there were further delays because of the late arrival of automatic regulators, the new system of power supply was brought into use just after Christmas.

At the end of 1953 we could derive some satisfaction in spite of the depth of the crisis in the spring and summer. The first major contractors, Whittals, had completed their tasks on the foundations and the power house and had cleared away from the site. We were left with a large ring of concrete and a tunnel leading through the ring beam to the central pivot housing, an enlarged power house with a new diesel capable of supplying far more power than we could utilize, and we had spent £80,000. However, alone on the site on Christmas Eve 1953 I could find little consolation in this progress which seemed inadequate to what would be necessary if the telescope was to meet the promised date of 1955.

14

Telescope control system and the control building

The blue book of 1951 contained a chapter on the control system of the telescope. Basically the telescope was to be built as an alt-azimuth instrument. That is, there were two separate motions: one by which the telescope could be rotated in azimuth through driving the bogies on the railway tracks, and the other the motion of the bowl only in elevation by driving through large racks at the tops of the towers. In principle the telescope could be used by simply driving the azimuth and elevation motions separately so that the bowl was directed to the required point in the sky. However, the actual motions necessary for many astronomical experiments are much more complicated. In particular it is an important requirement that a telescope can be driven in sidereal motion; that is so that if it is directed at a particular star, then it can continue to follow that star automatically, which effectively implies that the telescope is driven simultaneously in both elevation and azimuth in order to compensate for the rotation of the earth.

The positions of the stars and galaxies are given in celestial coordinates which are the equivalent on the celestial sphere of latitude and longitude on earth. The equivalent celestial coordinates are Right Ascension and Declination. The actual azimuth and elevation at which the telescope must be directed in order to point at a star of certain 'RA and Dec' can be calculated by spherical trigonometry. In practice it was necessary to do this calculation continuously so that the elevation and azimuth of the telescope could be constantly adjusted in order that the direction of the telescope should stay at the desired RA and Dec.

A system which enabled this to be done by using magslip resolvers was devised by J. G. Davies in the years when we were giving detailed thought to the design of the telescope. He constructed a bench prototype at Jodrell Bank in which the electrical output from the magslips was proportional to the sine and cosine of an angle. By using these in a servo loop system it was possible to solve the fundamental equations of spherical trigonometry with sufficient accuracy to enable the correct information to be fed continuously into the azimuth and elevation driving motors of the telescope.

The control system which we had in mind was the use of this basic analogue system in order to drive the telescope both in RA and Dec, and also in similar coordinates which are often used to specify the positions of stars in the local Milky Way system, in galactic coordinates or galactic latitude and longitude. Moreover, since we envisaged that the telescope would often be used to study the sun, moon, and planets it was necessary to arrange for an automatic parallax correction to be applied when desired.

The general design of a control system to do this, together with various other control features such as automatic scanning operations around a chosen position, had been drawn up in this way and before submitting the estimates for the 1951 application Husband had already approached several firms with a view to transferring the ideas into an actual system capable of controlling the telescope. When it became possible to proceed with the telescope in the spring of 1952 Husband was able to say that he had already obtained three tenders and that submitted by Dunford & Elliott of Sheffield was by far the lowest of the three.

In this way we began a long association with the firm of Dunford & Elliott of Sheffield, and became acquainted with Herman Lindars. It was soon clear that Lindars had many interests and that refined instrumentation of the type in which we were interested was only one activity of his firm. Furthermore it was soon realized amongst those of us who were musically inclined that this Herman Lindars was indeed the same person who was the eminent guest conductor of the Hallé and Royal Philharmonic orchestras.

The telescope brought many troubles into my life. It was also to bring unforeseen and unexpected pleasures. Lindars with the entire Royal Philharmonic at Jodrell Bank the morning after he had conducted *The Rite of Spring* from memory is one of those ineradicable treasures.

But in 1952 that moment was far in the future.

21 May 1952. Davies reported on visit to Husband and Lindars—Davies will provide revised and detailed specifications of the instrument panel as soon as possible, with suggestions as to those parts which should be built in the first instance as a pilot model. Lindars will then submit drawings of the proposed layout of the panel for the complete instrument. The pilot model will be built in this panel and will be used as part of the final design.

Of course the layout of this control system was intimately wrapped up with the design of the control building itself which we had agreed to site in the SE corner of Field 132 just far enough away to give the operator an overall view of the telescope. Husband's original drawing for this building was based on the idea that there would be two short spurs of a few rooms from the central control block and a sum of £16,000 was included in the estimates. However, we soon began to feel the need for more laboratory and administration space near the telescope.

12 Sept 1952 . . . Hayes[1] came over to discuss details of the control building. The main alterations which we have to suggest in the original plan are the inclusion of a lecture room instead of two of the rooms marked laboratories and for an arrangement for offices and also reception cum tea room.

There were further discussions with the architect in October and on 15 October 'apart from one or two minor changes we gave our approval'.

Lindars had tackled the job of the control system with great enthusiasm, and on 23 October Davies and I saw his proposed layout and computing arrangements—the demonstration being divided between the factory and his home. 'Money appears to be no object and it is quite clear that he will make no profit out of his present estimates for the computing systems.' At that meeting we settled on the basic layout with the essential controls at a console with the various indicating dials of the telescope position and of the celestial coordinates in racks to the right and left of the console. However, a few days later I had 'a spasm of worry about many of the details' and after a session with Davies sent four foolscap pages of notes directly to Lindars with a request for another meeting. This instantly resulted in a reprimand from Husband's office for dealing directly with a firm involved in the telescope construction. We still had much to learn!

Although we had agreed the final details of the control building in October, the next stage was long in developing. In December I was assured by Hayes that they were nearly ready to go out to tender and on 23 January 1953: 'On the control building Hayes said the bill of quantities was almost finished and that it should be ready to go out to tender within a week or so. He alarmed me by expressing the view that the cost of this building would probably be in the region of £25,000.'

This was already £9,000 more than the figure discussed on 12 September when it was 'still considered that the building as such could be constructed for £16,000', and came at a moment when the rapid build-up to the overall financial crisis was developing. Hayes had scarcely left my office when the call from the Bursar's office about the inconsistency in the steel prices came through (see page 49).

[1] W. I. Hayes, one of Husband's senior partners.

Feb. 17, 1953 . . . letter from Husband stating that he had looked carefully into the final details of the Control Building and was now quite sure that this would cost about £10,000 more than the amount in the building fund, which he named as £17,000. As a matter of fact the D.S.I.R.'s estimate only includes £15,000 for this.

He wanted to know if he should go out to tender or cut off a wing to bring down the price. 'I replied without hesitation that we should go out to tender with the building as it was and analysed why the discrepancy had arisen.' My analytical justification was based on the fact that the original sum which we had put in the memorandum for the building was £19,000, that it had been cut without justification to £15,000, and that the new estimate was little more than the increase in costs since my original figure of £19,000 was entered. Four tenders for the building arrived in mid April, and the lowest was £26,645, and it was this figure which was included in the revised estimate of £459,896 which we took to D.S.I.R. in April and May (see Chapter 12).

April 28, 1953. On Saturday Rainford phoned to say he had just received the plans of the control building, and considered that it was altogether too lavish. He again reiterated all his arguments about lack of thought . . . [On the Monday morning I went to his office.] . . . we discussed all the difficulties and increases in costs of the programme. As far as the control building is concerned Rainford was, I think, impressed by the consideration that the major part of the cost of this was in the central portion of the control room itself; and I made the point that the main increase in cost was due to the need for the basement for cable entry. I said that if pressed, we would reluctantly decrease the size of the wings by 20 ft each to restore it to the original lengths which were shown in the drawing in the possession of the D.S.I.R.

A month later, as described in Chapter 12, I was proposing to Rainford that we should delay the placing of the contract for the building so that he would feel that there was money in reserve to justify proceeding with the major steel contract. To this plan he agreed, and when the protracted negotiations on the steel contract were settled in mid July, 'Rainford confirmed that the steel contract situation was now in order and seemed a little less worried about the financial situation than previously. However he still refused to entertain any idea of placing the contract for the control building.' In spite of this, for reasons which are not at all clear, but probably because I was bothering him daily about the contract, authority was given on 9 September for Husband to proceed with the control building.

Unfortunately, Whittals, who had offered the lowest price for the building, now dragged their heels and on 24 September we were informed that 'after investigating the local labour position they had found that due to demands made by other large contracts in the area, they estimated the additional costs involved would be £1,050'. It is a pleasure to record that this manoeuvre to increase the cost was defeated because Husband interested another contractor in the work—Z. & W. Wade of Whaley Bridge—and their tender, which was a little lower than any of the original four, was accepted by mid October, almost exactly a year after we had settled the final details of the building. Even then there was another month of trouble with ditches and drains, and not for the first or last time I sometimes wondered if there was anything a university professor of radio astronomy was supposed not to deal with.

In the event the delays on this building were of little consequence, but at that time we were still talking of 1955 as a date for the completion of the telescope. By March 1954 the control building was not even showing above ground level.

March 3. The ice has now disappeared sufficiently for the asphalting of the basement to begin and about half of this was completed by teatime. . . . *April 12* . . . the basement walls are now up to ground level. . . . *May 6.* There has been some troublesome correspondence with Rainford over the minor issue of £1,000 which we have to find to run the auxiliary cable from our ring main to the control building . . . he caused me to be sent a rather irritable letter stating the position was extremely unsatisfactory, and 'how many more odd pieces of equipment etc. had been overlooked in preparing the original estimate'.

In reply I agreed that the position was most unsatisfactory and blamed a firm with whom Rainford and I were already at cross purposes (see Chapter 16). 'This morning Cannon[1] said that they had accepted this explanation and were sorry that the need for it had arisen.'

During the summer the brickwork rose steadily:

Sept. 19 . . . The control building moves lethargically forward, although the concreting of the N wing roof was completed last week. Goodall[2] now says Christmas for completion. . . .

Oct. 19 . . . there is clearly not the slightest hope of this building being finished before the end of the year. . . . *Oct. 27* . . . Goodall now thinks it might be February before we are in the building.

Towards the end of the year there was trouble about the electrical work in this building. Husband sent an estimate of £2,000 to Rainford whereas only £900 had been allowed. It was suggested that the balance

[1] D. G. H. Cannon, Chief Administrative Officer in the Bursar's department, who subsequently became Bursar of the University of York and later of Bristol.

[2] P. D. Goodall, M.I.C.E., Husband's Resident Engineer and now an Associate in the firm. Goodall worked imperturbably and with high efficiency on the site through all the buffetings of these years.

be taken from the telescope cable contract, but as the Bursar was quick to point out this would leave only £2,000 for the entire telescope cabling and he wanted an assurance that this reserve would be sufficient. This assurance could not be given, for by that time we were deep in our second financial crisis (see Chapters 17, 18). However, early in January the Bursar agreed to proceed on condition that the excess charges on the electrical work should be shown as an over-expenditure on the control building and not taken from the cabling contract. It was, in any case, an extremely minor matter compared with the financial horrors that were assailing us at that time.

So we moved slowly and painfully forward through all the stages of cabling, colours of paint, glazing, and heating.

Feb. 4 1955. Progress on the control building has been almost imperceptible. . . . *Feb. 9* . . . we have also been having blitzes on various aspects of the control building and T [one of Wade's men] assures us that we can get it in six weeks' time. This is almost impossible to believe since the plasterers are not yet at work, and the electricians faced with a £3000 contract have one youth and a boy on the job. . . . *March 31st* . . . The control building is plodding along slowly but there is little chance that we will be in it within the next two months. . . . *May 5* . . . moves forward shockingly slowly and at this stage one is reminded of the firm promises one had that the building would be ready by last October. In actual fact the people are only just now ready to start with the floors.

At last in late June and July the decorators moved in and the long process ended on 25 August when a symposium held under the auspices of the International Astronomical Union had its meetings in the lecture room of the new building and during September we took complete possession.

The miserably slow progress of this building, in the end, turned out to be quite inconsequential. During most of its building we were buoying ourselves with hopes that the telescope would be usable in 1955. In fact when we entered the building in the late summer of that year, the telescope was still a primitive steel skeleton and as disaster followed disaster on the financial front the chances of ever finishing the instrument receded.

Meanwhile, Lindars and his staff proceeded on their efficient way with the manufacture of the control system. In 1954 we had inspected the prototypes and passed the final drawings. When the International Symposium assembled in August 1955 we were even able to interest our visitors in two of the control racks and a mock-up of the console in the control room. There is little more to add about this part of the telescope. Another two years passed during which this complex apparatus was manufactured while the endless financial storms threatening the telescope continued. Then at the end of 1956 and early in 1957 first of all the indicating racks and on 11 April 1957 the control console arrived at Jodrell Bank and were wired up in their final positions in the control room.

15

The central pivot and the azimuth bogies

BEFORE returning to the main story of the telescope as exemplified by the steelwork and major financial issues, it is desirable to mention two other relatively small but vitally important items: the central pivot and the twelve bogies on which the telescope rotates. Our astronomical requirements presented the engineers with a difficulty since we asked that both the azimuth and elevation movements should be continuously variable from 2 degrees per hour to 36 degrees per minute, and that a control accuracy of a few minutes of arc should be maintained up to speeds of 4 degrees per minute.

Our maximum speed was really determined by the need to be able to rotate the telescope completely in a reasonable time—10 minutes—and did not in itself present any unusual problems since it corresponded to an angular speed of only 1 m.p.h. on the track with no particular requirements about control accuracy. On the other hand our slowest speed requirements were another matter. The lowest speeds were essential for the astronomical work and had to be at the highest accuracy. They were only one thousandth of the maximum and corresponded to a movement on the track of only one quarter inch per minute. The engineers faced an entirely unsolved problem in moving a structure of thousands of tons with such accuracy at this low speed.

Husband faced not only the problem of carrying the dead weight of over 2,000 tons but also the large variations in bearing pressure caused by the heavy wind loads on the structure. His final solution was to distribute the loading over a central pivot and a number of multi-wheeled bogies on the railway track. His design allowed the loading on the elevation trunnions to be transmitted directly to the tracks through the bogies. The locating pivot at the centre was designed to take the entire horizontal wind forces on the telescope in any direction and a vertical loading of 200 tons, sufficient to include a share of the dead weight of the diametral cross girder and the driving machinery which was housed in it.

The design allowed for about one half of the dead weight of the telescope to be taken on four driving bogies, two located directly under each of the vertical towers, and this was considered to be sufficient to provide enough track adhesion to prevent any wheel slip. The remaining load was to be taken by eight undriven 'wind bogies', two located at the foot of each of the large raking members which gave the instrument stability against overturning under maximum wind loading conditions.

The central pivot and the twelve bogies were clearly essential features of the telescope which had to be available at an early stage before much progress could be made with the erection of the steel superstructure. In the summer of 1952 Husband approached Stothart & Pitt of Bath and hoped that they would be able to manufacture the bogies. Our special requirements did not ease Husband's problem as regards the pivot.

Oct 16, 1952—discussed with Betts[1] the problem of getting the main power cable from the underground duct to the moving structure through the centre pivot. To our surprise he said he was designing slip-rings, this is of course forbidden in the specification and caused us some difficulty. He said it would be impossible to turn the main cable, even over a length of 20 or 40 feet, through 360°. After discussion we suggested that the best method would be to split the phases in the underground chamber and run four separate leads of smaller diameter through the central pivot.

Oct 21—Letter from Husband about the hole through the central pivot. He considers that the final decision must be to provide a 12 inch hole through the centre, and he also states that he is devising a method to take up the ± 200° twist on the main power cables.

The cable twister which was devised over the main pivot eventually carried many other service and research cables from the tunnel to the moving structure.

Oct 23—Husband has finished the drawings of the bogies and is trying to persuade Stothart & Pitt to make them. The trouble is that they are non-standard and engineering firms are reluctant to manufacture them.

Oct 31. Hayes said that Husband had persuaded Stothart & Pitt to manufacture the bogies but that their quotation was still awaited. He had also managed to interest a firm in manufacturing the central pivot . . . discussion also about the lab which it now appears necessary to build over and around the pivot and the new idea is to use this to house not only the cable connections but also the servo generators and distribution switch gear. . . .

Dec 10. Husband has now finalised the details of the bogies, the central pivot, the method of getting the cables from the annular laboratory to the rotating

[1] Frank Betts, M.I.Mech.E., a mechanical engineer with Husband's, closely concerned with the telescope design.

structure and the motor house which is to be built on the lower girder of the telescope. . . . Betts is going to Stothart & Pitt in Bath to settle some details of the bogie wheels.

On 18 December we had satisfactory reports of this visit and also of discussions with Hoffman's about the central pivot, but on 16 January 1953

. . . two disturbing pieces of news—Stothart & Pitt in Bath had still not tendered a quotation for the bogie wheels, neither had Hoffman's for the centre bearing. . . . *Jan. 28* still no quotation [but expected in a few days]. *Feb. 4* . . . he understood it was promised for next week . . . *Feb. 12* . . . letter from Husband containing a rather worrying piece of news that the price submitted by Stothart & Pitt for bogies was considerably more than allowed in the estimates and that he was now negotiating with the Yorkshire Engineering Co. who seem to be more interested in the job.

Actually Stothart & Pitt's price for the bogies was £35,000, which was more than half of the £60,000 which we had allowed for the entire driving mechanism.

Indeed we were moving rapidly into the first of the financial crises which has already been mentioned so frequently. In the 20 February meeting (see page 49) when I described the overall result as 'absolutely shattering', Husband had increased the estimate for the driving system by 32 per cent. (£19,000), '. . . a good deal of this is due to the fact that the original design of the central pivot was unsatisfactory, and whereas the allowance was £5,700, the final design which Husband brought over today will require £12,000'.

No one was in any frame of mind to place orders for the bogies or central pivot while the protracted financial and steel contract negotiations of that spring and summer were in progress. In fact, it was mid July before any new tenders were available for the bogies. In the revised estimates which had then received the approval of the D.S.I.R. we had allowed £79,000 for the driving system and the bogies were included under this heading. On 24 July we heard from Husband that he had received estimates from four firms ranging from £31,613 to £43,164. The lowest was from Thornton's of Huddersfield but we were agitated because their delivery date was a year and it seemed unlikely that the United Steel Structural Company Ltd. would be willing to commence the steel erection without the bogies.

July 24. Suggested to Husband that they could erect on dummies but he pointed out that distances of one hundredth of an inch are involved and although it may be possible to depress and elevate the huge structure, it would be a different matter to carry out any lateral movement . . . he hopes to get in a report on the central pivot in the very near future. In any case the lowest price for the bogies is only £1,613 over the £30,000 estimated and it is hoped there will be little delay in giving instructions for this work to proceed.

July 28. Rainford said the documents on the bogies had been sent on to the D.S.I.R., but again got in a complaint about there being an overexpenditure on an estimate which was prepared so recently. I am afraid that the next report on the central pivot and the trunnions is going to cause a major upheaval. I learned from Betts today that the price from Davy & United for the castings was in the neighbourhood of £12,000. This is the total amount which has been allowed under this heading—but it also has to include stub axles for the trunnions which amount to £7,500 and the bearings for the central pivot for which we have a quotation from Hoffmans for £4,000. It therefore looks as though this item is going to be overexpended by nearly £10,000.

However, with these troubles still ahead and ill defined we had the D.S.I.R.'s authority to proceed with the bogies early in August. '*August 7.* All that I now hope is that this [the control building contract] will be done before Husband's new missile arrives in the shape of the quotations for the central pivot and bearings which I am sure will represent a further considerable over-expenditure.'

By September Husband had negotiated a price of £2,475 with Davy & United for the castings and assembly of the pivot, and the total with various bolts and bearings came to £3,952.

Since this is less than £5000 the authority of D.S.I.R. is not required and I was relieved to hear this morning from Rainford that authority was being given for this work to proceed. The centre pivot has caused some concern since when Husband was here in the middle of August he expressed doubts as to whether the United Steel Structural Co. would be willing to put up the steelwork unless they had their centre accurately located. . . . this expenditure of nearly £4,000 leaves only £8,000 for the stub axles and trunnion bearings on the elevation movement. My impression is that the cost of these will be almost double this amount. [Sept 9]

However, some comfort could be had since 'all contracts have now been placed to enable the steel work to rise at least as far as the trunnions'. Alas, there was some contractual trouble. '*Sept 18*—unless the clause is rejected . . . the whole erection programme might be scuppered.—I am convinced after talking to Husband that although the absence of the bogies may not delay the construction it would seem extremely unlikely that the construction can begin without the central pivot. The delivery date given by Davy & United is May–June next year, which is already too late.'

Fortunately there was no delay of consequence.

Oct 13. Holmshaw[1] said the castings were now being made . . . and that the whole pivot stood a good chance of being erected at Jodrell Bank early in the new year, and he dismissed the idea that it would not be completed until the summer . . . full of optimism that the steel work of the telescope would be erected next summer.

Dec 1. Husband reported a stroke of good luck with Davy & United—they asked him to do an urgent job for them and he agreed on the condition that they would oblige by bringing forward the date for the central pivot.

January 26, 1954. One good item of news from Betts that Davy & United have cast the central pivot and were on with machining. Also that Hoffmans were having some trouble with the ball-races and had asked if Davy & United could accept a slightly larger size. This had been agreed and he thought that the centre pivot might be available in about a month's time.

Like most other things the delivery of the pivot was delayed, and also in common with the other delays on these miscellaneous items it really did not matter. By the time it came to the site on 11 May, the cranes were not finally erected (see Chapter 17), and the hopes of the major steelwork being erected in 1954 and the telescope in use in 1955 were really flights of fancy. However, the arrival of the pivot cheered us considerably. At last a bit of real machinery had been delivered. It was beautifully made and we filled ourselves with pride by demonstrating that we could rotate it with the push of a hand and yet it was to carry around the huge structure. Indeed, this device, which as Husband said had cost as much as an expensive car, was soon to be hidden underneath a mass of steel.

The D.S.I.R. press office had arranged for the press to visit the telescope site on 10 June. This was arranged when there was reason to believe that good progress would by that time have been made with the steel erection. In the event we were faced with a display of a railway track on a concrete ring, the shell of a building, and the central pivot. In an effort to make a minor improvement Thornton's were pressed to deliver one of the driving bogies, which they did on 9 June, but minus its gears. The press day was disastrous. Even in summer weather sixty press men assembled from all over the country might have been slightly peeved when they found nothing but partially assembled cranes. In fact it rained heavily all day and the entire place was a quagmire. Our press relations had got off to a poor start and as will be related in due course many years were to elapse before we were viewed with much favour.

In any case the bogie, after its wet outing from Huddersfield, had to be returned and it was mid July before a bogie complete with its

[1] T. Holmshaw, M.I.Struct.E., Husband's senior structural engineering assistant for over twenty years.

gearing was safely placed on the railway track. Then there was anxiety because at last steel began to arrive (see Chapters 17 and 20) and by August it seemed probable that the erection would be stopped because the diametral girder would reach the track before the second bogie arrived. This came, just in time, early in September. The delays in the bogie deliveries did not, in fact, interfere with the programme, the first of the wind bogies arrived in the middle of that month as the first of the chord girders was reaching the railway track. There were, of course, inevitably phases of apprehension:

> *Oct. 13*—no further bogies have arrived. Messrs Thornton's could only use men of particularly high skill on the machine which is handling the bogie steelwork. He [Husband] is now discussing the possibility of asking them to deliver the two driving bogies for the northern section before the wind bogies for the southern section. These will obviously be required during the next fortnight, if as expected the steel for the girders arrives on the site at the end of this week.

By 19 October the northern section of the girder was within a bay of the railway track but 'the two driving bogies for the northern side are promised in a fortnight's time'. '*Nov. 26*—one more bogie has arrived —a driving bogie for the north side—but this still leaves seven to be delivered and it will be difficult to start any erection of the tower until these have been delivered.' By mid December we had all the driving bogies, but the completion of the delivery of the remaining undriven bogies took a long time. '*Feb. 15 1955*—Thornton's brought over a bogie from Huddersfield this morning and having crossed the Peak they got into trouble at Bomish Lane entrance. However, the bogie was unloaded this afternoon and now there are only three more to come. Goodall says these are promised by the end of March.' In fact the final bogie was delivered in mid-April, but any satisfaction I might have felt at the completion of another contract for a vital part of the telescope was at that moment submerged under a mountain of other anxieties.

16

The driving motors

VARIOUS points about the requirements on the motion of the telescope have been mentioned in the two previous chapters and at this stage it is convenient to introduce the problem of the motive power. On page 29 I described how Husband and I visited the Admiralty's Gunnery Establishment at Teddington in 1950. Husband was impressed with the mechanical perfection of the mechanism used to rotate the 15-in. gun turrets and eventually purchased at a bargain price two complete 27-ft. diameter internal racks and pinions from the 15-in. gun turrets of H.M.S. *Royal Sovereign* and H.M.S. *Revenge* which were in the breaker's yard at Inverkeithing. Husband had of course considered other means of driving the bowl in elevation, such as by a large diameter rack system behind it but the cost of manufacture of racks of this size would have been prohibitive.

It was clear during this visit to Teddington that many of the automatic control problems of the driving motions of the telescope had been met in much more severe forms in these large gun turrets where there were problems of high acceleration involved—at least one problem which was not important in the telescope. We had therefore no hesitation in accepting the advice of the experts that the ideal solution to our problems would be to drive the elevation racks and the azimuth bogies through a servo loop system, and they particularly recommended the servo arrangement manufactured by a certain Company.[1] In essence the system we required was one in which the a.c. power from our diesel generators would drive rotary converters which would feed infinitely variable direct current to motors mounted on the bogies, and on the tower tops driving through the necessary gear reduction to the racks and pinions. The variability of the drive was of course to be determined by the output of the telescope control system described in Chapter 14. There were no outstanding requirements about the power. Husband estimated that even under extreme conditions of wind and ice the

[1] In view of our unfortunate dealings with this important British firm I have accepted legal advice to withhold their name from this book. They will be referred to throughout as 'The Company' and their system as the 'servo system'.

telescope could be moved by 100-h.p. motors and that normally the power required in each motion would be merely a fraction of a horsepower. He therefore decided that the rotary converters should be placed in a motor room immediately over the central pivot and cable turner, and that these should feed four d.c. drive motors each in azimuth and elevation. For the azimuth motion the four motors were to be distributed on the four driving bogies immediately under the towers and in elevation two motors were to drive two pinions on each of the racks.

In principle one of the most straightforward aspects of the telescope was to turn into a nightmare in practice. Following the advice of the Admiralty experts we had early and encouraging meetings with the appropriate department of the Company and when the proposals for the telescope went to D.S.I.R. in 1951 Husband's schedule included a sum of £28,600 which was The Company's estimate of the cost of supplying and installing the servo system and the associated gearboxes and electrical control equipment. In the revised estimates of early 1952 when the grant was finally amended this item remained at £28,600 out of a total estimate for all the driving system which had increased from £49,700 to £60,070. In the meeting on 29 April 1952 with the D.S.I.R. officials it was decided to proceed with The Company on the basis of this estimate. Husband was not in a position to arrange a formal contract straight away until the details of the azimuth bogies and their manufacturers had been settled, and little further progress had been made with that firm by the end of 1952. On 23 January 1953 my notes record that The Company was phoning Husband to inquire when progress could be made, but the bogie details were then still unsettled.

In the spring and summer of 1953, when we had to return to D.S.I.R. for another £104,000, the revised estimate of the driving system of £79,000 included an amount of £30,000 for the bogies, but we had no reason to revise The Company's estimate of the servo system which stood at £28,600. £12,000 was included for the main bearings and castings and £10,000 for the cabling. The difficulties of obtaining this extra money have been described in Chapter 12 together with the protracted sequence of events which finally led to the placing of the main steel contract in July. Alas, within days of the official letter of grant arriving from the D.S.I.R. I began to view the estimates for the driving system with dismay. The bogie situation was not too bad—the lowest tender from Thornton's was only £1,613 over the £30,000 allowed in the estimate. But it was at this stage, on 28 July (see page 69), that I wrote that Husband's 'next report on the central pivot and the trunnions is going to cause a major upheaval'. In September, when the price of the pivot had been negotiated with Davy & United, we had only £8,000 remaining for the stub axles and trunnion bearings on the elevation movement and we had already spent £5,000 of the cabling estimates on the main feed cable. Husband had already warned us in July that the stub axles alone would probably cost £7,500 and apart from any variations in The Company's servo system costs we were obviously in for a considerable over-expenditure. However, at that moment this was in the future and our energies and thoughts were directed to placing all urgent contracts so that the steel erection could begin.

Actually, although we had no inkling of this at that time, the real crisis here was to come from The Company. It was long in developing. The price of £28,600 as their estimate had stood for two years, and when Husband began finalizing arrangements with them after the decisions on the bogie manufacturers had been made, The Company began to suggest alterations.

Sept. 11, 1953—The Company have now suggested that standard traction motors be used instead of servo driving motors and this is delaying the placing of the contract for the bogies. . . . *Oct. 6.* The price for the trunnion bearings would be in shortly and he named a figure of about £9000. The situation with The Company also seems to be finalizing and the documents have now been returned to them for the issue of a formal contract. . . . *Oct. 12.* Husband recommends the acceptance of the tender for the trunnion bearings from Cooper Roller Bearings for £9,950. Unfortunately this brings the expenditure on the main bearings and castings up to £13,900 and only £12,000 was allowed in the revised estimate. Moreover this is not all. The figure does not include the stub axles which we already know will cost four or five thousand pounds. Rainford has sent on the documents to D.S.I.R. and has complained to Husband about overexpenditure and expressed the hope that something will be saved on The Company's contract.

This indeed was a vain hope. Husband was already getting worried about The Company situation and told me that he thought it unlikely that any money would be saved there. His worries were soon to be justified.

Oct 13 . . . letter from Husband containing the grave news that he believed The Company's estimate would be in the region of £40,000 to £45,000 instead of the £28,600 originally specified. He concluded that their original estimate of £28,600 was quite unrealistic, even allowing for the 30% increase in price which they claim has taken place. He had asked them to provide separate estimates for the driving motors and gearing, which were absolutely essential to move the telescope, and a separate quotation for the electronic arrangements necessary to give automatic motion. I replied, stating that in my opinion it was out of the question that we could obtain a further £15,000 or £17,000 on this contract, especially as we were already overspending considerably on the trunnions and bearings. It would be necessary to make an additional application which would be viewed with grave disapproval, but I do not think it is the slightest use doing this until we can have some indication of the cost of the mesh.

In the event, even Husband's feelings about the costs which he gave me in October were optimistic. On 1 December he brought to me the letter which had at last arrived from The Company in which they proposed three alternative systems with prices ranging from £50,000 to £55,000—and their initial estimate on which we had based our applications was £28,600.

The only bright spot about this is the deliveries which are offered, namely the motors and generators in a year, the servo system in fifteen months, starting equipment in fifteen months and electronic equipment in nineteen months. What can be done about this at the moment is shrouded in obscurity —I think it might be advisable to see Rainford and explore the possibility of a gesture from The Company.

The connections between The Company and the University of Manchester had been close and friendly over many years and Rainford readily agreed to explore the possibility of asking them to reduce the price or make a gift of some of the equipment. On 21 December I went with Rainford to see one of their directors. He argued that we must have changed our specification for The Company's price to go from £28,600 to £55,000 in two years. Fortunately we had the original letters which Husband had sent us for the occasion from which it was quite evident that the £28,600 was an irresponsible guess and even omitted several important items which were now priced in the new estimate. Rainford and I came away feeling that The Company was cornered and full of anticipation that our troubles would be dealt with by a friendly arrangement.

In all our dealings over the telescope Rainford and I were never, individually or collectively, more erroneous in our judgement. The director was unable to do anything.

Jan 6, 1954 Rainford phoned to say that he had heard from the director who reported that an investigation of their quotation had shown this to be perfectly in order and that there was no excess profit which his firm could forego. His explanation of the discrepancy was that the original one was merely a rough estimate. . . . *Jan 27* Rainford and I met the director. The result of his investigations and actions have been disastrous. First of all the approaches made to The Company were effectively blank, they had gone carefully into the estimates and satisfied themselves that they were reasonable. . . . Worse was to follow because the director then announced that he had met Lockspeiser [the Secretary of the D.S.I.R.] in his Club and had explained the situation to him. Lockspeiser's reaction was sharp and instantaneous—Manchester would not get another penny out of D.S.I.R. and must find the over expenditure from their own resources.

Rainford and I were much alarmed and distressed particularly at this unfortunate conversation which the director had with Lockspeiser. I suggested to Rainford that since we were clearly not going to get any more money then we should use what we had to buy the elevation drive from The Company and use tractors to pull the telescope round on the railway track. Later I wrote:

I realize that we are likely to be faced with an appalling situation should the cost of the mesh also come out 100% above the estimated value. The instrument will be useful without the driving system but will be absolutely useless without the mesh. It is therefore possible that we may be wise to withhold the driving system contract altogether and hold the money in reserve to cover any contingency on the mesh. In any case it seems that we shall have to place contracts with The Company for about £12,000 of elevation driving gears and axles even if we were going to adjust this by a winch. If we stopped at that point it would leave us with a safeguard of about £14,000 to cover overexpenditure on the mesh.

Husband was on his way back from Washington so I put these thoughts in a letter to him and also said that it was most desirable to obtain some big erection on the site as soon as possible. 'Many people are still taking every opportunity of saying the project must be cancelled.' On the Sunday following this (31 January) Husband drove through the snow from Sheffield to see me at home and we spent a long time discussing the problem. We agreed that Husband would seek an alternative source of supply—particularly for the gearing and miscellaneous items which amounted to four-fifths of The Company's estimate. He said that engineering firms were beginning to get short of work and he was receiving repeated inquiries for extra work. In any case we were extremely dissatisfied with The Company's delivery times even if they could be relied upon. He also agreed to investigate the possibility of picking up some cheap traction motors with which we could drive the telescope in a 'stop start' fashion provided the gearing was manufactured. Husband did not think much of my idea of pulling it around in azimuth by tractors.

He made the very good point that we had already spent £100,000 on an arrangement to make the thing rotatable and it was therefore a grave error of principle to give up the idea of automatic rotation, even as a temporary measure.

Feb 11. Greenall[1] and Cawley[2] here from D.S.I.R. Rainford also for a part of the morning and had a vigorous discussion with them on the general subject of over expenditure. . . . Cawley repeatedly emphasized that D.S.I.R. had no further money whatsoever in their next 5 year programme with which to assist the radio telescope—Rainford was reluctant to accept that this was the case especially for covering contingencies of increased costs arising from labour charges etc. The net result was that Cawley advised Rainford to raise the question at the very highest level with The Company.

1954 had opened with our major problems centred on this driving system expenditure. Before many months had gone we were deep in troubles of such magnitude that this particular episode of the driving mechanism receded from the front line of our anxieties. Our subsequent relations with The Company and the story of how the telescope was eventually driven is contained within the succeeding crises of the next few years.

[1] P. D. Greenall, Principal Scientific Officer, later Senior Principal Scientific Officer, 1952–61, in the grants division at D.S.I.R. Headquarters.

[2] Sir Charles Cawley, C.B.E., was then director of the division at D.S.I.R. concerned with university grants. In 1959 he was appointed Chief Scientist, Ministry of Power.

17

The crisis of 1954

In Chapter 12 I explained the tortuous path which eventually led in July 1953 to the authority for us to place the main steel contract with the United Steel Structural Co. of Scunthorpe. We had survived a financial crisis and had obtained a 33 per cent. addition to our original grant and now had authority to spend just under £460,000 on the whole project. The original estimate of the steel tonnage of 959 tons had gone up to 1,177 tons and so had the cost of the steel. At least by the end of that month all parties seemed satisfied and there was a mood of optimism that the steelwork erection would be completed in 1954.

The optimism was ephemeral. Within weeks, as we began to face some of the miscellaneous problems described in the last three chapters, it seemed impossible that the telescope could be made to work at that price. Large inroads were made into the allocation headed 'driving system', and we entered the major trouble with the driving system itself described in Chapter 16. In fact, almost within weeks of the decision to give the extra money, I could have expressed my doubts to Rainford and the D.S.I.R. The facts were not available. Neither I nor Husband knew then the troubles that were hidden in the driving system and the steel. Other vital possible uses of the telescope were coming to my notice apart from its use as an astronomical telescope. Every lorry that entered Jodrell Bank at that time was, for me, carrying a dream of the future. The path to the realization seemed to lie through an impenetrable jungle and the only way to keep it open was to fight for every item, however minor, with which we were ready to proceed.

As our troubles began to realize themselves in the need to place contracts costing ever more money I rationalized the situation by thinking of means of saving money on other parts of the instrument. At one stage this even reached the state of suggesting that we give up its full steerability, at least temporarily, in order to enable the steelwork to be finished. Indeed, if the economy had been stable, and if the best estimates which it was possible to make at that stage on so many unknowns had been correct, then the major confrontations of the future would have been avoided. In fact, every possible feature worked against us and led to a sequence of traumatic experiences which have left their mark on the individuals involved.

Even as I write about those days, ten years later, I cannot resolve the personal issues involved. I manoeuvred and pleaded with every individual I knew who might be in any position to assist. The difficulties were appalling and the eventual success of the telescope has not, in the deepest sense, erased the doubts about my personal solution of the problem which was thrown up by the events which I describe now.

The erection of the cranes

In the beginning all appeared to be going smoothly. On 11 September 1953 I recorded: 'Husband is expecting the receipt of shop drawings for checking in the near future. He still expresses complete confidence that the erection of the main steelwork will take place during next summer. . . .' On that day I wrote to Rainford answering a request to let him have my personal views on the probable overall timetable. It is reproduced here as a summary of the situation as it appeared to those of us at Jodrell Bank at that time.

11th September, 1953.

Dear Rainford,

In response to your request I set out below my comments on the probable progress of the Radio Telescope.

(1) *Foundations* (Whittals). Complete except for part of the tunnel and miscellaneous earthing arrangements. Most of this remaining work should be finished by the end of October.

(2) *Power House* (Whittals, Mirrlees, Brush). Building work complete except for finish of floor, which is awaiting completion of electrical work. Generator installed. Switchboards due August, now promised end of September. Whole should be complete 4 to 8 weeks after delivery of switchboards. Cable laid.

(3) *Railway track* (T. W. Ward). Delivery and fixing expected before end of 1953.

(4) *Bogies* (Thornton). Official delivery date, one year, but pressure will be brought to improve on this especially if the United Steel Structural Co. refuse to erect without (see below).

(5) *Centre Pivot* (Davy & United, Hoffman). Delivery expected April 1953. This delivery may determine date of steel erection (see below).

(6) *Steelwork* (United Steel Structural Co.). At a meeting held in May 1953 the following programme was promised:—

(a) The United Steel Structural Co. agreed to commence delivery of steel in February 1954 and complete all deliveries by May 1954.

(b) Cozens (Erection Contractor) agreed to start erection of gabbards and cranes in early January 1954, and erection of steelwork in March 1954. *Completion was promised by September 1954.*

My personal opinion of this programme is as follows:—

(i) there seems to be no reason why the United Steel Structural Co. should not keep to their delivery dates;
(ii) the erection may be delayed unless the centre pivot and bogies are available. There is a difference of opinion between Husband and Cozens as to whether the steelwork should be erected before the pivot and bogies are in position. My own opinion is that the centre pivot is essential before much erection takes place, otherwise the framework cannot be located accurately. On the other hand the bogies do not seem to be essential since no weight would be put on them in any case during the process of erection. If my view is correct, erection is unlikely to commence until the pivot is delivered, which would delay the erection programme by a month on present dates.

In a recent conversation Husband expressed complete confidence that the main steelwork would be erected in the summer of 1954.

(7) *Trunnion bearings and stub axles.* Contract not yet placed. Will be required before steel bowl can be built.
(8) *The servo system* (The Company). Quotation expected in 3–4 weeks. Will be required after erection of steelwork, that is in the autumn of 1954, but this is probably hopeless optimism.
(9) *Control Building* (Whittals). Completion unlikely until autumn 1954.
(10) *Instrumentation* (Dunford and Elliott). Not required until (8) available. No delivery difficulties anticipated—say spring 1955.
(11) *Mesh.* Contract not yet placed. Required autumn 1954–spring 1955 on basis of above programme.

From the above it will be seen that the most probable general programme is (a) completion of steelwork erection by autumn 1954,
(b) completion of entire instrument by summer 1955.

Yours sincerely,
A. C. B. Lovell.

During the autumn of 1953 the steelwork arrangements were proceeding normally as far as we knew. There were regular assurances that the erection of the cranes to handle the steel would start in January 1954 and with the firm assurances that the steel erection would be complete before the end of the year we were still thinking in terms of having the telescope in use in 1955.

25 November. Husband confirmed that the cranes were now working on a job which was scheduled for completion at the end of the year and that they were next assigned to Jodrell. . . . *15 Dec.* Holmshaw said that yesterday he was at Scunthorpe and that they were beginning to make good progress with the shop drawings and other arrangements for the steel.

Early in January 1954 Husband discussed the erection procedure with the United Steel Structural Co. and Cozens, the erection contractor. The first major problem was the installation of the cranes to handle the steel. Two cranes capable of movement on their own railway tracks were to be employed. About 1,000 tons of ballast had to be put down for these tracks. First a 120-ft. crane was to be erected and this crane was to be used to erect the 220-ft. cranes required for the main erection. It was then proposed to start the steelwork erection from the central pivot moving outwards to the railway track. The two towers would then be erected, and the main back girder of the bowl would be built on scaffolding from the diametral base girder of the telescope and the ground. It was then intended to invert this back girder and build the bowl underneath it. The construction of the bowl was to turn out to be an entirely different matter; in ways which at that time were quite unforeseen.

On 26 January 1954 there was a site meeting at Jodrell Bank with the erection firm of Cozens. Arrangements were made about the crane tracks and their foundations, power supplies to the cranes, and a mass of detail.

The erection programme seems to be determined by the ability of the United Steel Structural Co. to get the components out of their workshops, and this in turn is determined by the extraction of the drawings from the draughtsmen. Their representatives confirmed that work was now going forward as fast as possible, but that it was unlikely that they could commence deliveries until the end of May or early June. Cozens requires about two months to erect his cranes so this means that the erection must start in about six weeks' time and it will probably require most of this for the ballast and power to be prepared on the site. . . . this means it will be at least eight months from the beginning of the crane work, presumed to be early March, before the instrument can be finished. They agreed that they were worried about this excursion into the winter and said they would have preferred to have been starting the lower girders now.

In retrospect this innocent optimism seems almost absurd but none of the experts had at that time mentioned, even if they had thought, that the task was far greater than envisaged in this kind of timetable.

Feb 22 . . . slow progress—only 400 tons of stone for the crane tracks so far delivered—this will have to be speeded up considerably if the site is to be ready to receive the cranes by the end of the week . . . *March 9*—by yesterday evening over 1000 tons of ballast had been brought in to provide the foundations to the crane tracks but even so the easterly set of tracks is still far from complete . . .

Even the ballast carried its financial problem:

March 15 the double crane track road on the east side of the perimeter is completed . . . about 1½ thousand tons of stone has gone into the eastern roads alone. Goodall estimates that the expenditure here will be well over £3000 and this we have not allowed for at all in the estimates . . . no sign of the cranes . . . Holmshaw said that some of the crane steelwork was loaded a week ago and this was presumably in transit. . . .

March 18 Husband was impatient like myself of the non arrival of the crane tracks . . . slightly depressing about the expected rate of progress and he now thinks that it is unlikely that the bowl can be built this year, he thinks that by the autumn the erection will have reached the top of the towers . . . on the other hand he feels that there are still grounds for hoping that we may be able to use it next year.

After some unsatisfactory telephone calls to Scunthorpe Goodall was dispatched there to find out what was happening on the cranes: '*March 20* Goodall has returned from Scunthorpe with rather definite and better news about the cranes. The erection people have promised to come on Tuesday to commence the erection of a tower with which they will put up the 40 ft crane gabbards . . . he had seen these gabbards and they were finished.'

Goodall also brought back reassuring news of progress on the actual steel for the telescope. At the time of his visit various items for the diametral girder were already in the fabrication shops and he thought that erection could begin in eight weeks' time.

At last on 23 March a foreman arrived from the United Steel Structural Co. followed on the 24th by a large amount of steel. A 60-ft. tower was erected on the 25th, ready to hoist an 80-ft. tower which in turn was to be used to erect the gabbards. A 40-ft. crane was to erect a 100-ft. crane which could then erect the second 100-ft. crane. By 1 April the three 40-ft. gabbards for the main crane had arrived and the special bogies for the cranes were on the site.

April 12 the first crane is still in process of erection. Unfortunately Scunthorpe have decided that the 40 ft gabbards are not tall enough and are to erect on top of these further 20 ft sections—now the 80 ft tower which is being used to erect this first crane is no longer tall enough to do its job.

First change in design, 1952

The decision to increase the height of the crane gabbards even while they were in process of erection was an alarming eruption into practice of various discussions which I had with Husband, and it is necessary at this stage to explain the background to this matter which was later to become an issue of great controversy.

The basis for the original design of the telescope has been described in Chapter 10. The instrument was intended to work down to a wavelength of 1 metre, for which a shape tolerance of $\pm$ 5 in. and a reflecting membrane formed of 2-in. square mesh was adequate. At the time these decisions were taken there had been no work of consequence in the reception of radio signals from space at shorter wavelengths. In 1951 news began to circulate that Ewen and Purcell at Harvard, Muller and Oort in Leiden, and Christiansen and Hindman in Australia had all nearly simultaneously succeeded in detecting the spectral line emission from neutral hydrogen gas in the galaxy on a wavelength of 21 cm.[1] The existence of radio emission on this wavelength had been predicted in 1945 by a young Dutch astronomer, van de Hulst,[2] who did his calculations under the oppression of the German occupation of his country. The discovery was of high importance because it was the only line emission known in the radio spectrum. That is, the radio emissions which had so far been received were over a broad band of wavelengths, but this 21 cm. emission was a 'line spectrum' and changes in the wavelength received on earth would lead to deductions about the velocities of the gas clouds in the Milky Way and of the neutral hydrogen content of the galaxies.

It did not require much foresight to understand that this discovery would alter the trend and emphasis of much of the future researches in radio astronomy, and we naturally sought for ways and means of making the telescope usable on this frequency. There were two major problems. First the shape tolerances of $\pm$ 5 in. were quite inadequate for work on this short wavelength. Secondly the mesh size of 2 in. was too open and radiation of this wavelength would leak through.

In the autumn of 1952 I asked Husband if we could change the size of the mesh on the central 100-ft. section of the bowl to 1 in. $\times$ 2 in. It seemed to us that this central section should have the necessary rigidity to give a useful performance equivalent to a 100-ft. aperture telescope on 21 cm. After preliminary talks I wrote on 16 September 1952 to ask Husband to change the specification of the mesh in this central section 'provided it did not alter our requirements on the steel structure itself'. '*September 25*. Husband said this was causing more difficulties than originally anticipated and he estimated that the drawings would be delayed by about 5 weeks . . . apparently he is worried about the wind loading on the main back girder which is already extremely heavy . . .' In fact, the extra tonnage which had to be put in the back girder and the towers to meet my request was partially responsible for the financial crisis of 1953 which developed as soon as we tried to place the steel contract and which has been described in Chapter 12. This

[1] The three sets of observations were published in the same issue of *Nature*: H. I. Ewen and E. M. Purcell, 'Radiation from galactic hydrogen at 1420 Mc/s', *Nature*, vol. 168, p. 356, 1951; C. A. Muller and J. H. Oort, 'The interstellar hydrogen line at 1420 Mc/s and an estimate of galactic rotation', ibid., p. 357; J. L. Pawsey, ibid., p. 358, announcing the Australian results of W. N. Christiansen and J. V. Hindman which were subsequently published in the *Australian Jr. of Sci. Res.*, vol. 5, p. 437, 1952.

[2] Professor H. C. van de Hulst of the Leiden Observatory.

issue had been resolved and the money had been included in the new grant by the time the steel contract was placed in July 1953 and is relevant only as background to the major design troubles of 1954 which were first epitomized in practice by the need to increase the height of the gabbards.

The second change in design, 1954

It is impossible for me to give a strict analysis of the various factors which led to the great upheavals of 1954. Even from my own point of view I cannot write freely of the developments because many associated matters remain either classified or confidential. Further, I could not presume to have the professional competence to write in any detail about the many engineering problems which were revealing themselves to Husband. However, the story of the telescope would remain incomprehensible without some indication of the background to the 1954 changes.

It will be obvious that at this stage, early in 1954, we were realizing that the new 1953 grant from the D.S.I.R. would be insufficient. In fact, we were not at that time worried about the cost of the steel, which as far as we knew was adequately covered in the contractual arrangements, but our financial difficulties seemed to be tied to the driving system and the mesh. Furthermore, it is evident from my account in the last chapter that the D.S.I.R. had made it abundantly clear that we were unlikely to get any more money from them, the door of the Nuffield Foundation was firmly closed, and the University had no funds which it could allocate to the capital cost of the telescope. I was therefore urgently looking for another source of funds. Like many other scientists I spent some of my time in the committee structure of external advisers to certain Ministries. During 1953 and early in 1954 it seemed to me that for various reasons, radio telescopes were likely to become of great importance in the world of the future. I was not lacking friends and ex-colleagues in the appropriate echelons of these Ministries and I put up various memoranda for discussion early in 1954. As a result I was asked informally to hold discussions with Husband to find out how much would be involved in still further stiffening the bowl of the telescope to make it suitable for the tasks in question.

Y asked for a more accurate assessment of the cost of the modification and also of the scientific assessment of the performance of the instrument. I agreed to discuss this with Husband and let him have a report immediately. I felt that this had been a very satisfactory afternoon and with Y's interest in these other matters I felt little doubt that we should get the necessary money without difficulty. . . . I phoned Husband and he agreed to come over to discuss these things on Monday.

My conversations with Husband took place on 1 March. He said immediately that my new requirements could be met by using fitted bolts, site drilling and fitting on parts of the bowl structure instead of the black bolts as specified so far. He said also that some more stiffening would be needed in the back girder, and the supporting towers and the aerial tower. We would also have to decrease the size of the mesh from 2-in. to ¾ in., but this would not involve any new windage problems because the existing design was made on the assumption that the 2 in. mesh would sometimes get thickly iced. After phone consultations with Holmshaw, Husband said that half of the 34,000 bolts in the structure would need to be replaced by the fitted bolts. The cost was estimated to be £20,000 for the bolts, £13,000 for the extra stiffening, and £13,000 for various other minor changes. This formed the basis of the new memorandum which I sent to Y. It was important, I said, that a quick decision should be made because I had been informed by Husband that it would be impossible to introduce these modifications after the end of April. In fact Husband told me that he would proceed with the drawing office work straight away and if it was finally decided not to proceed with the modifications then only a relatively small drawing office cost would be involved.

It was clearly my duty to acquaint the other parties involved in the telescope with this new development if only in outline. I lost no time in dealing with this. On the same day that I dispatched the memorandum to Y, I wrote to Cawley at the D.S.I.R. explaining what was likely to happen. On 4 March I went to see Rainford in the University who took me to Mansfield Cooper[1] who was then acting Vice-Chancellor during Sir John Stopford's absence on leave. In my experience these two officers of the University have been unswerving in their support of progress. On that day and on many subsequent occasions in other matters I have approached them with schemes and plans apprehensive of some rebuff, or at least a cautionary warning. The doubts have always proved unnecessary. 'I was extremely pleased with their reaction which was that they would give all possible help as far as the University was concerned and that provided D.S.I.R. had no objection I could take it I was engaging in this affair with the University's full support and backing.'

With only two months to the deadline I wanted and expected an immediate assurance from the Ministry—after all the ground had been well prepared. Alas, the delicate threads were snapped by a

[1] Professor Sir William Mansfield Cooper, LL.M., LL.D., Professor of Industrial Law in the University, succeeded Sir John Stopford as Vice-Chancellor in October 1956.

departmental transfer early in March. On 4 March all was well, but on 11 March

> I was fed up with the protracted silence and phoned Y to enquire what he was doing about the memorandum . . . he said the thing was not looking so good . . . this did not seem to be at all in the spirit of the previous recommendations and I completely failed to comprehend why . . . with such a vital interest they should be reluctant to spend such a relatively small amount of money.

I failed absolutely to retie the vital thread. The original supporters either dropped away or retired into the safeguards of their official positions. The endless phone calls, visits, and pleadings led to no result other than an informal letter of regret from Y in mid April and more official communications of the decisions later in the month.

I was disillusioned and aghast at this blindness to the needs of the future. In the autumn of 1957 Britain had the only instrument in the world capable of giving non-visual information about the device with which the Russians launched Sputnik I. The fact that it had this ability owed little to those whose job it really was to look after these matters. However, it must not be thought that I bore any grudge. On the contrary, at the end of the affair, I was gratified that we had been left as free agents, to use the telescope on these matters as we thought fit and not shackled with a financial commitment which might well have ended with a grievous diversion of the telescope from its astronomical researches.

Practically in April 1954 I had no alternative but to tell Rainford that this road had closed and on 26 April I wrote to Husband stating that I 'could not authorise him to engage in any modifications to the steel work which would increase the cost over the present estimate'.

On the day of that letter

> Goodall informed me that the reason for making the gabbards higher was because the final cranes had to be made higher in order to carry out a different process of erection which had been necessitated by my scheme to alter the rigidity of the bowl. I had not realised that in order to do this Husband had increased the depth of the back girder, and this made it impossible to carry out the erection as originally planned.

I was filled with alarm—my request to Husband to abandon the scheme for increasing the rigidity of the bowl had no effect. It was too late, the drawings had been issued. I immediately crossed the Pennines with Davies and spent the whole of 28 April with Husband. In the earlier conversations Husband thought he could meet my requirements by inserting fitted bolts at a cost of £46,000. Now he said that the redesign of the girder was a cheaper and better method of achieving the same tolerance and rigidity and would involve only £11,000 on extra steel and about £4,000 on site drilling and fitted bolts.

April 29 (*referring to April 28*) I was in a dilemma but Husband was determined that he would not again change the design of the structure. . . . this is a far better design and since it represents only 10% of the cost of the steel work it is in any case within the limits with which he is able to predict the total tonnage of steel. . . . later when I queried Holmshaw he said that the estimate of 1100 tons on which the present price is based is unlikely to be correct to within 10% . . . his personal opinion was that it may well be a 10% over estimate since they had been generous in the allocation of tonnage.

We were by that time studying the problem of the reflecting mesh and obviously we had underestimated the cost of that. 'Husband also passed over some estimates for the cladding for the elevated towers and motor room. The estimate was for £3490 . . . this is an extra item for which no allowance has been made in the estimates, and since it seems extremely unlikely that we shall be able to buy the servo system there seems to be no reason for worrying about itat the moment.'

Cawley wrote from the D.S.I.R. to say how sorry he was to hear of the decision and asked if anything could be done now to facilitate a changeover at a later date. The only hope I held out in reply was that we might find it possible to install the smaller sized mesh within the financial provision already made for that item.

The summer of 1954

As far as I was concerned the activities described in the last few pages had ended by Husband deepening the back main girder of the bowl at a cost which was within the limits of which he was able to estimate the overall tonnage at that stage. There was a worrying ripple early in May when on a visit to Scunthorpe Husband was told by the United Steel Structural Co. that they had already cut £2,000 worth of steel for the original back girder which they thought it would be difficult to dispose of; he said also that they were making difficulties about other minor changes and the increase in height of the gabbards, but that he was visiting them again next week and that these were minor matters on which he had no doubt that they would meet us. 'I told Husband that my great concern was how to escape from the net which would close when the bills came in' (4 May).

I was excessively bothered. I could not see how we could drive the telescope without spending another £20,000 or £30,000. The major steel structure on which we had believed everything to be finally settled was taking an indefinably ugly financial look, and miscellaneous minor items—ballast for the cranes, handrailing on the walkways, cladding for the motor rooms—were all beginning to add up to significant sums which as far as I could see were not really in the estimates.

May 6. Last night when driving home I realized quite suddenly that we could save a good deal of money on the telescope by dispensing with the access tower. Our original requirement was that we should be able to get to the focus with great ease, but in view of the great financial troubles and because we have now decided to put a good deal of our apparatus at the base of the tower it seemed to me perfectly clear that we could save between £10,000 and £20,000 by omitting this tower entirely.

I phoned Husband and after much discussion:

I think he is really pleased with the idea because he said they have taken 18 months and have so far failed to design a satisfactory structure. It is certainly the most dangerous part of the telescope from the users' point of view. . . . Husband thought we might save £5000 but my estimate is that we shall in fact save much more, £5000 represents the cost of the steel but to this would have to be added considerable sums for electrical cables and controls.

That was the end of the access tower and we had at last, on paper at least, saved some money. Simultaneously I was pressing Husband on the cost of the cladding for the motor rooms on the diametral girder and on the tower tops. 'I suggested that the cost of these represented a fair sized house and that it might be cheaper to take the system off the rotating structure into a brick building.' I did not win that point and Husband said that in his view the cost of this cladding was chargeable to the driving system contract.

At least in that case we were not committing money which we did not have. The remaining funds in that account were quite inadequate to buy The Company's servo system as already explained and we were then intending to buy cheap traction motors so that the surplus remaining could, in principle, go to these housings.

Meanwhile on the site of the telescope itself the erection of the gabbards and cranes was proceeding with maddening slowness. '*May 11* . . . most unfortunate delays in the erection of the first crane. The latest trouble is that one gang of men have had to be dismissed because they were asking for more money in order to work in the country . . . the foreman now thinks that it will take the rest of the week to have this first crane in operation.' In fact, it was not working until, 'after a further agonising series of delays', the middle of the following week. '*May 19.* The jib is 120 ft on the gabbards which are already 60 ft above the ground, hence the total height of the jib at 180 ft is almost exactly the height of the trunnion bearings of the final instrument. This afternoon it unloaded several sections of the gabbards of the second crane.' This was not one of the final erection cranes. The plan was to use this to erect the second crane with gabbards at 120 ft. The first crane would then be dismantled and the second crane would build crane No. 3 and the diametral girder of the telescope simultaneously. 'I wrote to Husband to complain about the very slow progress and pointed out that every single promise which I had ever been given had been broken' (May 19).

An official of the United Steel Structural Co. came to a lecture which I gave in London on 21 May and 'he assured me that it was still his intention and hope that the telescope erection should be finished this year'. In retrospect it was clear that any competent assessment of the situation made nonsense of such a timetable. In any case the schedule of erection was falling behind hopelessly. '*June 1st* . . . irritating delays on the erection of the second crane due to the fact that the intermediate gabbards have not been finished in Scunthorpe—there is little chance that this second crane can be working for two or three weeks.' Goodall was a frequent visitor to Scunthorpe and often came back with disquieting news which led to more calls and letters from me to Husband about the progress. On 1 June Goodall said that he had actually seen some of the base girders and had got a promise that they would be delivered in two or three weeks, 'hence it seems reasonable to assume that the actual erection really will begin in June. This is indeed small comfort after the expectation that the whole of the base girder would be erected by this time.'

Husband himself remained cautiously optimistic. On 3 June 'he still thinks that the instrument will be at trunnion level by September. I gather that there is very little detailing remaining in Husband's office and that the final drawings are all issued and checked apart from some of the purlin details on the bowl.' On that day two of the three gabbard sections of the crane were up to 120 ft., and as the hope of a beginning to the erection of the diametral girder in June was fading rapidly I once more wrote an agitated letter to Husband. This eventually elicited a reply from the General Manager of the United Steel Structural Co. who said that my remarks were 'disturbing to those who were doing their best to get the job moving'. The implication of his outline of the immediate programme was most depressing. He gave 26 June as the target for completing No. 2 crane and said they hoped to assemble the third crane 'before their works holiday began at the end of July'. There was clearly no hope of a start on the erection until August. A discussion with Goodall on the basis of what he had seen in the shops during another visit to Scunthorpe led me to conclude 'there is little likelihood, even if everything goes according to plan, that the structure can reach the top of the towers before the end of November. The programme is wildly different from everything on which we have so far based our hopes.'

At last by mid July both of the final erection cranes were complete and a 20-ton girder assembly to be bolted to the central pivot as the mid section of the diametral girder had been delivered. On 10 August, after days of delay due to the loss of packing pieces, this large girder assembly was placed on the pivot and the erection of the steelwork of the telescope had begun.

The third change in design, 1954

Through May, June, and July there was always the nagging background anxiety of the driving system. Husband and his team approached several firms to attempt to find someone who would give a competitive price for the gearing assemblies. If that could be settled, then arrangements might be made to find some traction motors at a bargain price so that we could move the telescope.

However, this and all my previous troubles vanished in the face of another cataclysm which occurred in August. Husband had been to Ceylon.

July 22. Husband appears to be catching up after his visit to Ceylon. . . . this morning there was an envelope packed with several letters from him. In at least two of these he referred to difficulties which he was having with the United Steel Structural Co. . . . He also referred to the worry which the increasing weight of the back girder was causing him.

I came out of a conference and spoke to him on the phone; '. . . he confessed he did not know the final weight to which the structure would attain. He said that the deepening of the girder had caused unexpected lengthening of various other members and it would still be some weeks before their weight estimates could be finalized . . . the position is altogether most unhappy.' The next day he came over to see me; '. . . his worry was not a fundamental one concerned with the structure itself but only with the cost of the increasing weight . . . there seems little to be done at the moment in any case. Husband said that it would still be some weeks before their final weight estimate would be available.'

On Saturday 14 August I had been away playing cricket. On my way home I called in at Jodrell. 'There was a letter from Husband to say that after several checks he had been forced to the conclusion that the original design of the bowl had become unreasonably heavy under the new tolerance conditions. He went on to say that in order to deal with this he was proposing to redesign the bowl and would let me have preliminary drawings in a few days.' He referred to our financial discussions in March and April as the top limit on our excess expenditure. But I had already made it perfectly clear at the end of April that the extra money for this stiffening would not be available! I summoned Hanbury Brown[1] and Davies to meet me on Sunday. We agreed that if we were ultimately faced with the dilemma we could, on the present estimates, save the budget by abandoning all aspects which were not absolutely necessary, such as any idea of driving it at all in azimuth, at least as a temporary measure. 'We all felt that a transit telescope capable of being preset by hand in a comparatively crude manner would give us many years of work.' Husband was away that weekend and on Monday I recorded that my conversation with him was not very satisfactory since he said they were hoping to save money by this redesign, and that the full investigations and detailing would take another three weeks.

Husband has given his own account[2] of the severe problems with which we had presented him and which had slowly but inevitably forced him to conclude that the huge back girder design would have to be abandoned:

> A preliminary assessment of the dead weight of the structure was particularly difficult at the outset; there was no precedent for a framework of this size mounted on trunnions and subjected to such stress variations on account of the rotation on the horizontal axis and the extremely variable nature of the wind loadings.
>
> The first design, which was worked out in detail, envisaged a heavy cradle-shaped girder between the trunnions, with radial cantilever members connected with a circumferential secondary system carrying the reflecting membrane or wire mesh. An erection scheme was worked out in detail by which the cradle girder was to be built on comparatively low scaffolding from the diametral girder joining the two trunnion towers, and then the cradle was to be inverted through 180° so that the radial cantilevers could be conveniently handled by the erection cranes. The whole bowl structure was to be completed in the form of an umbrella, the only additional scaffolding being that necessary to steady the structure against rotation.
>
> As the stress analysis proceeded it became apparent that while the

[1] Professor R. Hanbury Brown, F.R.S., who joined me in 1949 to study for a higher degree having already had a distinguished career in military and industrial radar. His brilliant developments in radio astronomy at Jodrell Bank kept our academic status at the highest level during these years of distraction. In 1964 he left Jodrell Bank for Sydney in order to use apparatus which he had developed for measuring the angular diameters of stars under favourable sky conditions.

[2] *Proc. Inst. Civ. Engs.*, vol. 9, p. 65, 1958.

deflexions at any angle of elevation would be within those originally specified for 1-m radio reception, the structure did not lend itself to providing the increased stiffness which was highly desirable if the telescope was to work efficiently on the 21-cm waveband. Stiffening up the cantilevers added weight, which was not a very serious matter when the bowl was directed at the zenith, but which produced troublesome torsional effects in the cradle girder when the bowl faced the horizon.

These deflexions and dead-weight difficulties led the Author to take more interest in the geometry of the paraboloid, and particularly of the nature of its deflexions during rotation on the elevation axis. When directed towards the horizon or at low angles of elevation the parts of the reflector near the upper and lower ends of a diameter at right angles to the axis of rotation tended to slump under the effect of gravity, with a corresponding displacement of the focus, which would upset the directional control of the radio beam. On the other hand, if a form of construction could be devised by which the periphery of the reflector could be kept reasonably in one plane and circular, inevitable deflexion distortions of the central portion of the paraboloid, due to the changing angle of gravitational forces during rotation, would mainly cause a small variation in focal length and a much smaller angular displacement of the radio beam. With a "rigid" periphery the reflecting surface, supported by main trusses arranged radially, even after deflexion, would still closely approximate to a paraboloid; small variations in focal length, if necessary when working on short wavelengths, would be compensated for by a relatively simple servo-mechanism.

With this general arrangement the bowl is deepest and the focal length minimum when the instrument is directed at the zenith, and shallowest when directed towards the horizon.

These considerations encouraged the adoption of a 'space-frame' form of construction with a very deep circumferential girder. Sixteen main radial trusses were used to connect the panels of the circumferential girder to a central hub. The calculation of stresses in the framework was tedious but reasonably determinate, bearing in mind that the whole structure was supported at two points only. The most difficult parts of the framework to design were the members connecting the trunnions to the circular structure. The shear forces in the framework due to dead weight and wind forces acting at all directions through 360° were troublesome to deal with, and required a much greater weight of steel than had been allowed for in the preliminary structural estimates.

The final design of the bowl structure was extremely tedious because of the large number of loading conditions which had to be considered, and because every alteration affected the position of the centre of gravity. It was obviously highly desirable that the centre of gravity of the whole of the bowl structure should be closely on the axis of rotation, and the necessary arithmetical calculations fully occupied an electric calculating machine for nearly a year.

The inquisitions to which I was subjected years later made it perfectly clear that I was guilty of a fatal error of judgement in not reporting this situation in detail immediately to Rainford and the D.S.I.R.

I have already made some remarks at the beginning of this chapter on the general situation which confronted me during those days. I do not excuse my failure to judge that at that moment in mid August I should have done things which I did not do. My explanation for inaction is firstly that the facts were not available. We did not know until Husband had completed his investigations whether the new design would be heavier than the back girder version. In spite of this I might well have talked to Rainford or the Vice-Chancellor about it, but at that moment Rainford was on holiday and the Vice-Chancellor was in hospital. Before they returned I went to the assembly of the International Scientific Radio Union in Holland and a whole month had elapsed before the normal threads of communication were restored. Even when I returned Husband was away but my great anxieties were to a certain extent calmed in a conversation with Holmshaw.

> *Sept. 11* . . . Husband away but spoke to Holmshaw who has sent me a general arrangement drawing of the new bowl. I am satisfied that the redesign which abolishes the back girder and puts all the steel in a cylindrical structure containing the paraboloid, is going to be much more rigid than the original design. . . . Holmshaw said that the weight was less than the old design.

This conversation restored my hopes that if there were excess costs then in an emergency they could be covered by abandoning the driving system.

Three days later Husband was at Jodrell and confirmed my tendency to inactivity by saying that the new bowl was only 520 tons, which was very little heavier than the original design, compared with the 700 tons which the redesigned back girder had reached before it was abandoned. At that moment it looked as though we would be overweight to a small extent because of an underestimate of the amount of stiffening required in the substructure. Indeed it seemed that the acute anxieties of August about the financial consequences of the redesign were unjustified.

As will be seen in due course we were eventually to be faced with a huge over-expenditure on the telescope. The public and private inquiries were to cause me much personal suffering, and as so often happens in these cases the situation was so complex that the inquiries tended to concentrate about an obvious point which is not necessarily the vital one. When the files and records were searched the investigators found this apparently alarming situation in which I had agreed privately to a 'change in design' with Husband. It is true that I had at that moment agreed to a change which, to the onlooker, altered the appearance of the bowl of the telescope, but to me as a scientist it was a modification enabling Husband to meet our design requirements which had been accepted in the 1953 revision. Years of experience have shown that in fact, it did just that and no more. Furthermore at this distance in time when other people have tried, and often failed, to build big telescopes, and incidentally have wasted vast sums in their efforts, it is clear that Husband was correct when he said that this revision of the bowl would save money. I agree that I was guilty of a technical administrative error of judgement in August 1954. However, it was not this error which led to the ultimate over-expenditure. This arose partly from issues which still had to be revealed and partly from the difficulties of costing the large part of the structure which was in no way affected by this business.

While acknowledging the error it is I think necessary to add that had I been primarily concerned with my own skin at that time then either there would have been an enormous delay in the telescope or Husband would have been forced to return to the back girder design and this would certainly have led eventually to an even greater financial upheaval.

18

The autumn of 1954

IN September of 1954 when my initial alarm about the bowl had subsided it seemed that Husband's idea would enable the telescope to be finished with a tonnage of steel not greatly in excess of the available figures on which the existing grant had been made. We had about £26,000 left in the driving system allocation which was quite inadequate to meet the new prices proposed by The Company for the servo system. As far as the viability of our overall financial position was concerned the condition seemed not unreasonable, since if called to account, this amount of money would cover any likely excess costs on the steel. At that moment Husband had no evidence to the contrary and hence during that autumn I returned to the attempts to bring in enough money from outside so that we could order the driving system and end up with a fully steerable telescope as originally intended.

In the previous chapter I mentioned my disillusionment in March and April of that year when certain channels of action closed. During the summer more and more information came to me which made me determined to try again. I did not hide the fact in these approaches that we needed more money in order to provide the driving mechanism to make the telescope steerable, and to cover our underestimates on the tonnage and cost of the steel.

After the immediate astonishment of the spring rebuff I retired into the affairs of the telescope and my own researches. I was awakened from my lethargy by a conversation with a colleague, as we were walking out into a London street after a meeting. When I received Husband's letter on 14 August I was already deep in this renewed activity and the new financial crisis which this seemed likely to open increased my resolve. In August and September I visited and was visited by a considerable number of people whose primary business lay in fields other than astronomy. By a strange irony Sir Ben Lockspeiser the Secretary of the D.S.I.R. had arranged to bring the Chairman of the Public Accounts Committee[1] to Jodrell Bank on 20 September. I shall relate in Chapter 26 that Sir Ben had already appeared before this

[1] Sir George Benson, the Labour M.P. for Chesterfield.

Committee on the affairs of the telescope and when I met them at Crewe Station his first words to me, in the presence of the Chairman, were that he had explained to the Committee that no further D.S.I.R. money would be available for the telescope. It was a bad start.

It is not uncommon for people to come to Jodrell Bank in one frame of mind and leave in another. At the end of the afternoon the Chairman of the P.A.C. left full of sympathy for us and our problems. However, he explained that this could not affect his public duty or that of his committee when faced with activities which exceeded the estimates. It did not. Fortunately I knew Lockspeiser well and since he had other duties in the neighbourhood he was spending the night in my home near Jodrell. The Secretaryship of the D.S.I.R. has attracted eminent scientists and Lockspeiser was no exception. He understood our problems from the inside, he was, too, a charming man and a most accomplished pianist. Under such circumstances the true issues on both sides readily break through the formalities of the public exchanges. Sir Ben asked me to arrange for a certain person who knew the other side of the story to come to his office and express his interest in having the telescope completed quickly and in a steerable condition. It was done and done quickly and efficiently. I had retied the thread which snapped early in March.

There was instant action on the part of the D.S.I.R. On 2 October Cawley wrote to me to say that as a result of Lockspeiser's visit to Jodrell Bank and the subsequent representations made to him, there had been 'further discussion of the position in the Department'. He asked for an explanation of why we were short of money and 'what additional sums of money would be necessary to fulfil the original plan'. Rainford was unwilling for a reply to be sent unless we had a written statement from The Company to confirm that their price for the driving system of £55,775 still held. Husband was away in Ceylon but Betts pressed The Company for this assurance which he did not get. On 8 October Betts phoned to say that The Company's reply had announced that 'due to a special wage award last April all prices in their quotation had now been increased by 5%'.

On this basis Rainford agreed that my reply to Cawley could be dispatched. The draft had already been settled in discussions with Rainford and Cannon and the final letter dated 8 October 1954 was a long three-page explanation of the current situation. On the subject of the driving system it said:

In the estimates £27,000 was allowed . . . for the cost of the servo system and gear boxes . . . but when the Consulting Engineers reached the stage of finalising the contract last autumn The Company gave a renewed price of

£55,775. . . . a certain amount could be accounted for by rising costs but by far the major discrepancy arose because of gross estimating errors on the part of the firm. [On the steel contract:] . . . the continuing rise in costs of material and labour has had significant effects on the renewed prices submitted to D.S.I.R. early in 1953. On the work already completed there has been an over-expenditure of £7,500 and there is bound to be a significant increase in the main structural steel price before the completion of the work . . .

In his letter Cawley had asked me 'how precisely are you planning to meet the situation if no additional funds are forthcoming? How would this affect the work you could do with the telescope?' I explained our scheme for abandoning the access tower and steerability so that it would be used only as a transit instrument, and using the money thereby saved to cover the increased costs of the steel. Finally I said the new price from The Company was £58,563 and

hence the original allocation of £27,000 would have to be increased by £31,563 in order that the contract could be placed. Our major reserve to cover the existing over-expenditure of £7,500 plus the appreciable increase which we feel is bound to occur in the main steelwork would then be removed. . . . we do not feel justified in making further guesses as to the probable increases in cost which are likely to occur and which, in any case, are largely out of our control. We think that this matter can only be assessed much nearer the completion of the instrument when any further savings in mesh etc. can be balanced against the present and future over-expenditure. . . .

In making this last statement I was following the advice of Lockspeiser who thought we should wait until next spring when we would know how much extra money we really wanted. However, it was impossible to do this if we were to drive the telescope and with the backing from the external sources I hoped that we could get immediate authority for the extra £31,563 required to place the servo system contract, pending a better assessment of the total requirements at a later date.

The case seemed to be overwhelmingly strong and I did not doubt that the authority would be given. I think that at that stage certain Ministries would willingly have given direct financial aid but it was well understood that the Treasury would frown on money entering the same project from different organizations and it was considered expedient to do it through the D.S.I.R. channel. 'Y said he had seen Lockspeiser and that he was certain we should be given the extra money needed to drive the telescope.'

Alas, goodwill and intentions, even from the highest, had to be processed through the complicated channels of a Government department and the attrition and the wear and tear began again. A week after my letter to Cawley it was being explained to me in London that the D.S.I.R. had 'committed all its money for the next five years' and 'why did the University not get help from local interests'. By mid October I was still confident but less sure of the outcome. 'My impression is that they (the D.S.I.R.) will find the extra money but that there will be the usual agony as it goes through the Advisory Council, the Treasury, etc. Cawley asked if six weeks was too long to wait . . .' On 18 October I wrote: 'This interim situation is therefore not unsatisfactory . . . the thing I most fear is a conditional kind of reply from D.S.I.R. to the effect that they will give us the extra money for the driving system but that the University must find the money to cover the over-expenditure due to price increases. This would certainly not be accepted.'

Fortunately Sir Hugh Beaver had become the Chairman of the D.S.I.R. Advisory Council and it is necessary to explain how he came to Jodrell Bank in the midst of these affairs.

19

Simon of Wythenshawe

It would be imprudent of me to have the presumption to write in any general terms of two of the most distinguished figures in Manchester's history—Lord and Lady Simon of Wythenshawe. However, over the whole period of the development of Jodrell Bank of which I write they had a vital influence. Lord Simon was Chairman of the University Council and Lady Simon a member of that body, and their various interests penetrated like a sword into a host of University developments.

My first direct confrontation with the Simons occurred at Jodrell Bank in the summer of 1946 when we were in the rudimentary state described in Chapter 2. It was the pleasant habit of the Professor of Botany to hold a small garden party when the strawberries were ripe in the grounds of his department at Jodrell. This occurred on a Sunday afternoon and the most eminent of the University hierarchy were invited. It was an occasion for the botanists, but in view of the existence of our trailers in his grounds the Professor deemed it advisable to suggest that my wife and I should also now attend this function. We did so with trepidation and were enjoying ourselves quietly in the shade of a tree at a respectful distance while the glories of the herbaceous border were being inspected, when we received an imperious summons. Lady Simon had spied a strange structure towering over the delphiniums in the distance. There was no escape, an inspection was demanded and we set out on the march over the rough ground to the next field but one.

How much land did I occupy?—A few square yards.—'Ernest do you hear that? Dr. Lovell has to get all this apparatus in a corner. He ought to have two hundred acres.' I was myself, at that time, thinking in terms of the odd acre. I had been judged and found wanting mainly in terms of the extent of my outlook—a new and electrifying experience for me at the age of thirty-three. At the end of that day I imagined that I was to be left to absorb this opinion. My imagination was lacking in knowledge of the Simons; I was to be helped. On the Monday morning at 9.30 a.m. I was summoned to the phone by a frantic Bursar (not

Rainford but Kaye his predecessor). 'Lovell, what the d— went on at Jodrell yesterday afternoon?'—'A tea party. Why?'—'Only that Lady Simon has already been in my office and made me get out all the maps of Jodrell Bank. She says you want two hundred acres.'

In 1954 when we were deep in our troubles we had the two hundred acres and more, but the interest of the Simons never waned. The annual summer afternoon tea parties of the botanists were supplemented by tea parties of a different kind—nearly always on a Sunday afternoon. The Simons would swoop with the suddenness and keenness of the eagle, bringing with them their weekend prey who was either to be educated or to be persuaded to help us. We would produce the tea sometimes in our home and sometimes in my office at Jodrell. The stature of each of the Simons and the breadth of their experience and knowledge was so enormous that the two of us together could not stand up to the questions and inquiries from either of them singly. One felt as though a computer had prepared the stiffest examination on earth for our personal benefit and that the answers would be fed back and assimilated for the future benefit of ourselves and others.

However, we passed and survived, even if without distinction, and the fact that we did so was a crucial factor in the history of the telescope. There were situations, which I still have to relate, in which Lord Simon would have been entirely justified in calling a halt to the construction of the telescope. I suspect that many Chairmen of University Councils would have done so without hesitation. Not Lord Simon. He simply adored big projects. As the governing director of the Simon Engineering Group of Companies he was not unacquainted with large enterprises or their difficulties, and although he did not mince his words in expressing his horror of the administrative muddle in which I had landed the project—'the trouble with you Lovell is that you're a rotten administrator'—he was determined that it would not fail.

As we fell deeper into our troubles so Lord Simon, both by virtue of his position as Chairman of Council and because of his personal enthusiasm, became more involved in the telescope and his name will occur often in this story. Even to the end I never overcame the apprehension accompanying his visits. However, the apprehension was of a friendly type arising from the deep inward desire of the junior to do well in the presence of the master. There was to be one dreadful occasion when the apprehension turned to a frustrated terror and in this, as will be seen in due course, not only Lord Simon but his entire cohort of directors were involved.

It was with this Sunday afternoon escort of the Simons that Sir

Hugh Beaver[1] occasionally visited Jodrell Bank. The first occasion on which he visited was, I think, in the midst of our 1954 autumn troubles and it was another of our strokes of fortune that he had just then become the Chairman of the Advisory Council of the D.S.I.R.

Sunday Oct 3 . . . in the afternoon Lord Simon came out with Sir Hugh Beaver and on the site we encountered the general manager of the United Steel Structural Co. [quite accidentally as far as I remember] . . . He [Beaver] seemed completely unconcerned about the financial aspect of the telescope and said that no engineering job of this magnitude had been built since the war without the price rising by several times.

[1] Sir Hugh Beaver, K.B.E., died in January 1967 at the age of 76. At the time of his visits to Jodrell Bank he was the managing director of Arthur Guinness & Son & Co. Amongst his many distinguished public services he was the Chairman of the Committee on Air Pollution, the report of which led to the Clean Air Act of 1956, and during the war he became Director-General of the Ministry of Works.

20

The telescope rises

ALTHOUGH Sir Hugh Beaver, Lockspeiser, and the senior officials of the D.S.I.R. were individually anxious to ease our problems at this stage they could but work through the official channels at their disposal. With our past unsatisfactory financial history on the telescope the machine was, perhaps, more sensitive than it might otherwise have been. My long letters to Cawley in mid October led to a visit of the Chief Finance Officer[1] who came to Jodrell Bank with Rainford on 28 October. He said that D.S.I.R. could not possibly instruct us to proceed with the expenditure of another £30,000 to cover the driving system costs unless they knew what the final overall picture was likely to be. In my letter to Cawley, quoted in Chapter 18, I had explained the difficulty of doing this, but it was no good; the official machine was forcing us once more into the kind of estimating which we wished to avoid. I was left with instructions to present a new table of costs after discussions with Husband.

I discussed this with Husband later in the day of this visit and we agreed to meet again in London after a few days during which he would assemble the appropriate figures. On 4 November we drafted a letter for Husband to send to Rainford. This gave the information that, as was already known, we wanted another £30,000 to proceed with the driving system. Altogether we asked for £50,500, the other major part of which was £15,000 for the additional steelwork. The explanatory comments which accompanied these various details seemed to me to present a reasonable case and I was not unhappy. Indeed one's state of happiness in this situation was relative to a constant state of torture—we had at that time already paid £165,000 of the £390,000 allocated for the project and we were still almost at ground level.

Rainford was greatly displeased with this letter: 'he now wants to know why if it was possible to give such a precise estimate in 1953 there is now this vague talk of increases in tonnage' [Nov 9]. We were still in a tangle of negotiations over the driving system, and, ever optimistic,

[1] S. H. Smith, Finance and Accounts Officer in D.S.I.R. Headquarters from 1950 onwards.

I was persuading myself that we might save on the £30,000 extra for which we had asked. I therefore suggested to Rainford that we should not, after all, communicate our new requirements to D.S.I.R. until we had the answer to that problem and could put forward more specific proposals. Rainford agreed and I wrote to Cawley to say that 'our delay in making an application for additional money was not on account of any lessening of the urgency but because we had been able to suggest an alternative source for the gearing which was the most urgent part of the programme'.

The unfortunate story of our relations with The Company on the servo system and driving motors has been told in Chapter 16. The impact of the crisis in the summer of 1954 first led to the idea of a partial abandonment of the motorized drive, followed by the resuscitation of the original plan with the still further increased price from The Company (see p. 97). Throughout the summer Husband had been investigating the possibilities of a cruder form of drive and of sources for the gearing associated with the driving motors so that we could break away from The Company's servo system around which all these various assemblies had so far been designed. On 26 September Husband came over to see me to say that he had just returned from Germany and thought that it would be possible to buy suitable gearboxes for the azimuth and elevation drives 'off the shelf', and also that Siemens might be able to provide an alternative drive system of the 'Ward-Leonard' type, much more cheaply and with far better delivery dates than The Company offered. I promised to find out in London if there were likely to be any political difficulties in purchasing items from Germany for inclusion in the telescope; at least it seemed that the offer of good delivery dates and competitive prices from abroad might radically alter The Company's attitude. The situation with The Company was certainly not improving. Apart from the further 5 per cent. increase in price (see p. 96), we had been informed by the manager of their Contracts Department in early October that 'owing to the fact that they had accepted very big contracts in the department it was now no longer possible for them to keep to their original delivery dates and that they would be unable to effect any delivery until September 1957'.

Oct 13. . . . This, of course is fantastic and I wrote to the director [see Chapter 16] asking him to find out if this was realistic because if so it meant that we could never place a contract with The Company even if the money was available.

In addition to his investigations in Germany, Husband had been exploring other sources for the supply of the gear assemblies in the U.K.

In mid October he had received a quotation from the Power Plant Co. of Middlesex. Although the price was not dissimilar from The Company's estimates for these items (about £15,000), the promised delivery date was altogether more realistic—six to eight months as against three years. We were, indeed, glad to have this information in our possession because, as a consequence of my letter to the director of The Company on 13 October, a meeting took place on 27 October to discuss the situation with The Company.

Betts tabled a schedule showing that all machinery and gearing would be required at various times between next February and the end of September when the telescope was expected to be in operation. The Company's representative said that this was impossible and they could not deliver under three years . . . I was particularly perturbed by their statement that at least 6 months would be taken up in development and redesign in the drawing office even if given over-riding priority . . . I pointed out that this was in direct contradiction to the statement made in 1949 when we were advised to use their standard servo system in order to save this redesign which they are now claiming would take so much time.

Eventually Betts produced the information that he could obtain the gear assemblies in six to eight months. With this bottleneck removed from the schedule '. . . they have now gone away and will let us know as soon as possible what their new delivery schedule would be and the price. I phoned the director and pointed out to him that this was now his opportunity to get The Company to present us not only with a favourable price but also with a good delivery date.'

These discussions had already taken place when Rainford and the Chief Finance Officer came to Jodrell on 28 October. The Power Plant quotations for the azimuth gearboxes were identical with those from The Company, but the elevation gearboxes were cheaper. In fact the amount involved was £22,000 against £24,000 in The Company's quotation. Neither Rainford nor Smith saw objections to placing this contract immediately and Husband was anxious to proceed although he was still hoping for a firm figure from Germany. The appropriate documents recommending that the contract be placed with Power Plant had already been sent to D.S.I.R. by the time Rainford received Husband's letter in early November and this was the point to which I made reference in writing to Cawley about the alternative source of the gearing.

A few days later Husband received the information from Germany for which we were waiting. The quotations for the gearboxes were only a few hundred pounds less than those of Power Plant and did not justify any change in our recommendation to D.S.I.R. On the other hand the news from Siemens was good—they thought they could produce the driving system for the telescope in nine months but needed more discussion before making firm proposals. Husband asked if I would go to Germany with him for these discussions but I suggested that J. G. Davies should go instead since he had made a detailed study of the driving requirements from our point of view.

Husband and Davies intended to fly to Germany on 22 November but the visit had to be postponed because Siemens were not ready. On the other hand Husband received from them on that day a quotation for a complete driving system at £17,200 with a delivery in nine months. With the Power Plant quotation for the gearing this gave a total of £39,000 for the whole system compared with the £58,000, three-year offer from The Company. What little hope and interest we had remaining in The Company was quickly evaporating. When Husband and Davies were finally in Germany on 6 December we had the revised official communication from The Company which they had promised on being released from the necessity to manufacture the gearing. They offered the servo system for £27,100. Not only was this £10,000 more than the Siemens price but the delivery promises remained hopeless, there being no promise of completion until 1957.

Davies and Husband returned from Siemens quite happy about the suitability of the proposed Ward-Leonard drive. However, their original estimate needed revision because some unnecessary items had been included and others, such as magnetic brakes and the repeater racks, omitted. The delay was unfortunate. It was essential that the question of a further grant for the telescope should be discussed at the meeting of the D.S.I.R. Advisory Council on 15 December and we had still not acted officially on the request of the finance officer to put forward our new requirements. I had many conversations with Cawley in D.S.I.R. during the days preceding this meeting and conveyed to him informally the essence of Husband's November letter to the Bursar—that we needed another £50,000. Even if we saved on the Siemens quotation our extra requirement would be at least £40,000.

It seemed to me that the autumn stimulus of Lockspeiser and Beaver was being lost in the mechanism of grant giving. Cawley was cautious in the extreme. He said that the question of a University contribution would certainly be raised again and advised me to ask Rainford to send him a note of the extent of the University's present financial involvement in Jodrell Bank.

Dec 13 . . . If the D.S.I.R. refuse to sanction the purchase of the machinery from Siemens or give us the necessary money to order the system from The Company then I said [to Cawley] that we should ask for permission to buy the driving motors at a cost of £7000 from The Company and drive the instrument in a stop-start fashion . . . if we do not place orders and make up our minds in the near future we shall neither have the gearing nor any of the motors by the time the steel work is finished.

I had insufficient faith in Lockspeiser and Beaver!

Dec 17 . . . Called on Cawley with some apprehension to hear his report on the reaction of the Advisory Council . . . to my amazement he said they had taken the very definite view that every effort should be made to finish off the telescope in its original fully steerable form and had instructed the Department to find ways and means of getting the extra £40,000 or £50,000 required. I had been prepared for gloomy news . . . it was with great relief that I learnt of this definite and progressive attitude.

Cawley was worried about where this money would come from and had again written to the Nuffield Foundation asking them to share this additional cost; but the situation there was not at all favourable.

My relief was only partial. No one was sure that this '£40,000 or £50,000' was really all we required. We still had no revised quotation from Siemens and on Christmas Eve 1954 as I surveyed the litter of steel on the site I noted that 'the future is dark with financial anxiety'.

Indeed the future was darker than we knew, but throughout all our troubles the framework of the telescope continued to grow. The beginning of the erection of the steelwork through the frustrating delays of the summer has been described in Chapter 17. On 10 August the large girder assembly at the centre of the diametral girder was placed over the centre pivot, and within days the telescope started to grow towards the railway tracks.

August 16—the steelworkers have completed the entire centre section of the base girder and have now run out of steel. Perry[1] said that though another load was on the way he doubted if it would arrive until the end of the week. . . . *August 18* . . . the steel for the next section of the cross girder still not delivered. In fact Perry told me that they were not expecting to finish it in the works before this evening . . . the cranes are in a state of inactivity; they have been for the past two days and are likely to remain so for the rest of the week . . . *August 21* . . . several hundred tons of steel are in various stages of transit by road and rail between Scunthorpe and here . . . the main deliveries are expected early next week and it is then anticipated that quite rapid progress can be made with the next two or three panels of the diametral girder.

Rapid progress was indeed made. When I returned after two weeks' absence in Holland at the International Scientific Radio Union the diametral girder had reached the railway track on the southern side and the chord girders on that side were going into position. By 20

[1] Douglas Perry, D.F.C., Site Engineer for the United Steel Structural Co. Ltd.

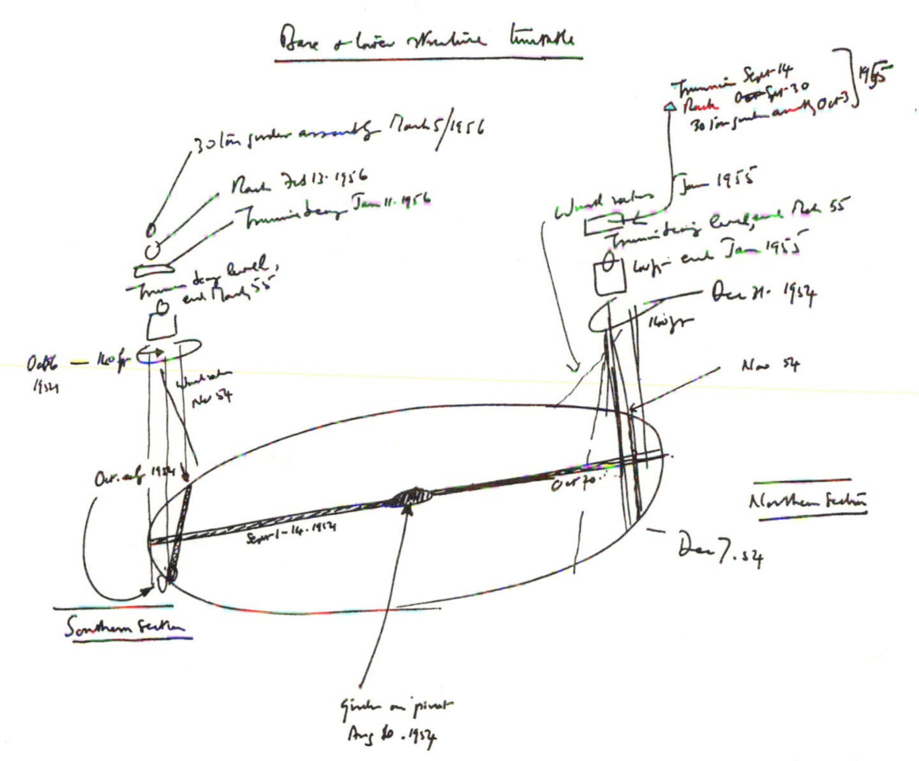

September the base girder work on this side was so far complete that one of the cranes was being moved back towards the pivot in order to begin the northern section. At the beginning of October the heavy steel sections at the ends of the chord girder on the south side were in place and the cranes began the erection of the first section of the southern tower.

Oct 6 . . . Spectacular progress continues to be made on the steel erection. At the end of last week the first two sections of the southern tower were in position; early this week the third section was erected, and today two splices of the fourth section have been placed in position. The height of the top of this section of the tower is now nearly 140 ft . . . *Oct 13*. The main progress has been on the extension of the diametral girder to the north. There has been some trouble with the supporting girders for the southern tower and this has not yet gone to a greater height. There is a good deal of levelling to be done but Goodall says that the steelworkers would much prefer to be erecting steel because of the extra payment which they get for working at a height.

A week later the diametral girder was out to the railway track on the northern side, but of course much detailed infilling of the structure and riveting had to be carried through. Indeed the riveters were to prove a constant source of trouble. '*Oct 29*. Husband was reasonably satisfied with the progress on the site although there has been a set-back because the riveters have left the site because they are dissatisfied with their wages . . .' They returned in mid November. Under winter conditions progress slowed up.

Nov 15 . . . steel erectors working on the wind rakers to the southern tower. Last week they erected most of the western section of the northern chord girder . . . *Nov 22* . . . the wind rakers on both sides of the southern tower now in position . . . some progress has also been made with the base of the northern tower . . . *Nov 26*. Hewson[1] [from Scunthorpe] confirmed that they were hoping to clear away from the works all the tower structure to the top by the end of this year. The most serious matter at the moment seems to be the delay in riveting. . . . Husband thinks that this will now be the holding feature in the construction since it will be impossible to make any progress with the erection of the bowl until the 17,000 rivets in the base structure have been attended to . . . *Dec 7* . . . still great trouble with the riveters and no work has been done for the past week.

The target for the completion of the tower structure was the end of January 1955 and discussions were already taking place about the erection of the scaffolding necessary to support the bowl during construction. The basic decisions were made at this time. '*Dec* 7. Last week there was a further meeting to discuss the mode of erection of the bowl at which it was decided to build up the scaffolding from the diametral girder to take the annular ring at the centre of the bowl and also a circular rim of scaffolding to take the outer perimeters.' We were also finalizing the details of the laboratory which was to swing underneath the bowl and of the approaches to it and the aerial tower. In this first week of December the chord girder on the northern side was complete and the tower on that side reached to the second section. Gales and floods continuously interrupted the erection, but on 14 December I wrote: 'this afternoon the erectors had got two of the girders of the northern tower to the 140 ft level'.

On Christmas eve when I noted the gloom in the financial outlook I also observed that

since last Christmas the site of the telescope has been transformed. Then there was nothing above ground, not even the railway track and Wades were only digging the drains of the control building. Now the steelwork of the two telescope towers rises 140 ft above the ground, and those and the cranes are an impressive sight for many miles. Even the control building has its roof on . . . however, there is one thing in common with last Christmas, the driving system is still in a state of no progress.

[1] Philip Hewson, Erection Manager, United Steel Structural Co. Ltd.

21

A quarter of a million pounds in debt

THE year 1955 opened quietly with little sign of the tempests which were to strike us. The telescope towers were at the 140-ft. level, the erectors were working on the northern wind rakers and we hoped that the scaffolding for the bowl would go up in the spring. We still had placed no orders for the gearing or the motors. Although we had authority to proceed with Power Plant in the autumn, Husband was unwilling to commit himself until the issue of the motors was settled because of uncertainty about the motor speed. We thought we needed £50,000, and in view of the Advisory Council's reaction in December we did not doubt that our grant would be increased as soon as we were in a position to submit our requirements which had now been held up since October.

Siemens versus The Company

Although we were poised to place the orders for the gearing with Power Plant and for the motors with Siemens we failed to make progress. At last on 20 January Husband said he had received a letter from Siemens: 'they are so anxious to be associated with the telescope that they are willing to include the magnetic braking without extra charge. Before submitting a final recommendation Husband is waiting for the German estimates for the gearing . . . he thinks that the German price may be cheaper than that of Power Plant with delivery immediately.' On 20 January Rainford saw Smith, the Finance Officer of D.S.I.R., in London, and told me on his return that Smith 'wanted our prices for the driving system as soon as possible but did not anticipate much difficulty, although because we were proposing to purchase from Siemens it would need Board of Trade approval'.

At the end of January Husband went to Ceylon where he was concerned with large public works and I continued my daily phone calls to Betts in his office; but always it was the same response—no further information from Germany. At last on 20 February, Husband, back from Ceylon, brought over a telegram from Siemens offering the gearing at £12,000. This seemed fine, but until we had the details we could not compare this with the Power Plant figure. By early March Husband had been able to obtain enough information to indicate that the overall cost of the German gear assemblies and bearings would be £5,000 less than the British quotations, and on 4 March he brought over the full analysis which in essence was that The Company's price for the gearing and motors was £56,000, the combined Power Plant gearing plus Siemens motors would be £42,000 and the German gearing plus German motors £38,000.

At last the situation seemed to be clarified. By placing the orders in Germany we would save £18,000 over The Company's costs and have delivery in nine to ten months instead of three years. The delays had made the placing of the contract an urgent matter. The constructional programme had receded badly, but the telescope towers were soaring upwards and all were optimistic that the bowl steelwork would go up during the summer and autumn. The D.S.I.R. had agreed in principle to let us have the money to buy the machinery and at last I thought we were moving out of a hopeless tangle. I was too innocent of the ways of the commercial world: we were moving into an even deeper tangle. On 7 March Rainford told me that Lord Simon had heard that we were proposing to place these orders with Siemens. He and the Treasurer, Sir Raymond Streat,[1] became greatly alarmed at this short-circuit of The Company issue, and Rainford said that Sir Raymond Streat had written privately to the Director of the parent organization of The Company to ask him if he was aware of this situation. Like myself, Rainford was exasperated by the dealings with The Company but he was afraid that questions might be asked in Parliament about the inability of British firms to supply machinery for the telescope. Under ordinary circumstances this might not have concerned us, but for reasons which will be described later the affairs of the telescope were becoming a matter of political and public concern and we were already in too much disfavour.

In spite of this it was our intention to make the official request to D.S.I.R. to place the contract with Siemens as soon as Husband provided Rainford with the necessary documents. Another fortnight elapsed and then on 22 March Husband came over with a very bad piece of news. There would be a 20 per cent. import duty on the Siemens system, putting us nearly back to The Company's figure. Rainford remained optimistic: he thought we might avoid the duty.

[1] Sir Raymond Streat, K.B.E., was at that time Chairman of the Cotton Board. He succeeded Lord Simon as Chairman of the University Council in 1957 and, as will be seen, played a cardinal role in the affairs of the telescope.

March 22. My own impression about the programme is that with good fortune the bowl will be built before the autumn . . . the delay over the placing of the orders for the driving system and gearing have again made me extremely gloomy about this matter and I think we shall be extremely lucky if next year at this time we are beginning to get delivery of the gearing.

An April of despair

There were many periods of despair during the years of construction of the telescope. April 1955 was one of the blackest. When the month opened we seemed on the point of placing the orders which would enable the telescope to be driven. Within a few days the likelihood of doing this in the foreseeable future vanished, and this owed little to problems of import duty or the ethics of using foreign machinery.

The favourable Siemens price renewed our hope that the extra '£40,000 or £50,000' which had been the basis of our talks with D.S.I.R. since the autumn would be sufficient to finish the telescope. Rainford and Cannon had arranged to meet Husband and me at Jodrell on 5 April with the hope that we might finally agree on the submission of the documents to D.S.I.R. for the German purchases. The first alarm sounded on the previous Saturday when Cannon read over the phone Husband's preliminary letter to them in which he said that 'in the next few days we would be discussing estimates for the driving system ranging from £65,000 to £76,000 and that the reasons for much of this expenditure could only be understood by experienced mechanical or civil engineers'.

Even on the top figure from The Company there was another £20,000 which had gone into this estimate. In fact, Husband had now assembled all the various miscellaneous costs necessary to complete the driving system, including the cabling. However, the real crisis did not spring from this sequence of underestimates. On the Monday morning Rainford smoothed out the thick pad of documents on his desk and pointed at one line: 'braking £15,300'.

I had known since a bleak Sunday afternoon in January that Husband would have to change the existing scheme for the elevation control but no price had emerged until this moment. The reasons for these new troubles were complicated and revolved around years of discussion about the reflecting surface of the telescope and the effect of high winds on the structure. Today, when the design of complex steel structures is amenable to analysis by computers, it is hard to remember that during the design stages of the telescope extremely little was known about many aspects which are now common knowledge. When the day of reckoning came on the telescope the inquisitors could see readily that our original design for a reflecting membrane of open mesh had been changed to a solid steel sheet reflector, and in their eyes I had reached a private agreement with Husband to change the design not only once but several times. Even after the events in the spring of 1954 described in Chapter 17 it was held that I had agreed to still further changes from a mesh to a solid skin reflector, which had led to further large items of over-expenditure.

In fact, the circumstances which led to the beginning of the 'braking' saga had far deeper origins than a simple change of mesh to a solid sheet. It is true that in the early years of the concept we envisaged an open mesh reflector. It is also true, and this point was always conveniently forgotten by those who searched for easy explanations, that one of Husband's first acts when he was presented with the design problems was to ask the National Physical Laboratory to carry out wind-tunnel tests on some samples of the mesh which we proposed to use. The results of these tests were given in an N.P.L. report in November 1950, and they included the case for 2-in. mesh in a condition which was expected to result in practice under blizzard conditions, that is, so much icing that it would become impervious to wind. Thus, even the initial calculations made by Husband assumed that under bad conditions the reflecting mesh would behave as a solid reflector and even, under extreme conditions, that the top half might be snowed up more quickly than the lower half, thereby worsening an already serious turning moment due to the differential wind pressures. The ultimate design change from mesh to steel sheet did not therefore, by itself, introduce the new crisis, although—most fortunately as it transpired—it illuminated the need for the modification to the elevation drive.

The original design called for a reflecting surface of 2-in. mesh. A change of the central 100-ft. section to a 1 in. × 2 in. mesh was made in 1952, and early in 1954 the special circumstances which led to the desire to use a smaller mesh over all the bowl has been described in Chapter 17. During these years we had considerable contact with Messrs. Greenings of Warrington who had machines capable of weaving mesh of the gauge and length which would be required. On 29 April 1954 I noted that Greenings could supply the 50,000 sq. ft. of mesh required for £13,500 in copper or £14,800 in a cadmium-copper alloy and that Husband desired to use the alloy because its tensile strength was 40 per cent. greater than the copper and its resistance only 50 per cent. greater. Although these figures were well inside the £20,000 allowed in the estimates no one had any reasonable suggestions as to how to make the thousands of joins needed which could only be done when the long strips in which the mesh would be supplied were in position on the bowl framework. At that moment we could only conclude that 'there seemed to be a good hope that we might be able to buy and fit the $1\frac{1}{8}$-in. mesh[1] within the £20,000'. As mentioned in Chapter 17 I wrote to Cawley at D.S.I.R. on 29 April to say that we were proposing to use mesh of this size.

During the summer D.S.I.R. proposed that we should use aluminium mesh, which according to Cawley would save us four or five thousand pounds. Greenings said they would weave the mesh in aluminium but were even more concerned about the joining problems than in the case of the copper. The problem of making the large numbers of joins in an economic way which gave not only good conductivity but sufficient structural rigidity was rapidly beginning to appear almost insoluble. In September Husband first raised the possibility of using a solid sheet reflector as a way out of these troubles. Our initial reaction was cautious for two reasons. First, there seemed to be no available information about the efficiency of riveted sheets as a reflector when the wavelength of operation was comparable with the size of the individual sheets. Second, although the solid membrane added no new problem to the ultimate safety of the telescope, since the calculations had already assumed complete icing of the mesh, there was the question whether the wind loading under more normal circumstances might not be greater with the solid skin to such an extent that the specified operating wind speeds for the instrument would have to be lowered. I wrote to Husband

> expressing our great interest in the solid skin provided he could guarantee good conductivity and provided operations in wind speeds of 30 m.p.h. would not have to be sacrificed. [Sept. 19] . . . Today [Sept. 21] Husband wrote again on the subject of the solid sheet reflector. He tried to reassure me over the question of wind speed, but in fact has failed to do so. We are analysing our records for the past year and hope to have a fuller appraisal of the situation in the very near future. Husband also asked for further clarification about the d.c. conductivity of the joins and we confirmed that a conductivity at least 100% as good as the material itself, was essential.

During that week J. G. Davies analysed our wind records to get an idea of what fraction of the time would be lost giving estimates at various wind speeds. 'It is clear that no decision can be taken on the solid sheet bowl until he [Husband] has completed his calculations on the deflections . . . I am not only concerned with the safety of the instrument but also with the distortion which the bowl is likely to suffer at much lower wind speeds . . .' (26 September).

[1] That is centre to centre. The $\frac{3}{4}$-in. mesh size mentioned on p. 85 was the clear aperture. The $1\frac{1}{8}$-in. size was the smallest we seemed likely to get within the £20,000 overall ceiling—Greenings' price for a $\frac{13}{16}$-in. mesh, for example, was £19,000 in copper or £21,000 in cadmium-copper, for the material alone.

As the need for a definite decision became urgent, so our thoughts tended increasingly to the idea of the solid skin. Husband's interest was structural, because of the additional stiffness over the mesh, and our interest was operational, because there would be no loss of signal by leakage through the mesh and no leakage of unwanted interfering signals through the mesh from behind the bowl. Further, in mid October, Greenings wrote to say that they could weld the aluminium mesh, but that the joints were very poor, structurally.

Oct. 28 . . . a number of minor points discussed with Husband, not the least of which was his idea that the membrane of the bowl should now be made of stainless steel. He pointed out that in the case of aluminium the expansion was such that buckling and corrugation would certainly result. Later on the idea of stainless steel was replaced by the idea of mild steel, which could be painted and which would be very much cheaper.

In order to clarify the new uncertainties about the effects of wind on a solid sheet instead of the mesh Husband again sought the advice and help of the National Physical Laboratory. On 4 November I lectured at the N.P.L. and afterwards visited Scruton[1] who was in charge of the aerodynamics division. In fact, he had recently completed, for an entirely different purpose, a wind-tunnel investigation on a solid membrane reflector 2 ft. in diameter. His report on this seemed to contain all the information which Husband might need, although the bowl was relatively somewhat shallower than the telescope bowl. Scruton agreed to carry out further tests immediately if the information in the report was insufficient. I gave the report to Husband in London the next day, and after a short study he decided that the scaling up from the reflector used in these tests was not entirely justified when applied to the deeper profile and different supporting arrangements of the telescope bowl. Consequently he arranged for Scruton to test a one hundredth scale model of the telescope in the wind tunnel.

The scale model was taken to N.P.L. early in December and Scruton started on the wind tunnel tests immediately. The data about the turning moments were in line with expectations—the most out-of-balance wind force occurring at about 60 degrees elevation, but of a magnitude well within the safety factors already allowed. Husband was most pleased with the cooperation which he obtained from Scruton and said on 20 January that he had now obtained nearly all the required information and would visit me in a few days to discuss the results.

The background drama of the telescope occurred as much in my home

[1] C. Scruton, a senior member of the staff of the Aerodynamics Division at the National Physical Laboratory and a leading world expert on the interaction of airflows with civil engineering structures.

as in my office. It is perhaps fortunate that I never formed a habit of sleeping away my Sunday afternoons. On 23 January 1955 the most agonizing of all Sunday afternoons still lay in the future, but this one was bad enough. Husband came to tell me the final news about Scruton's tests. Up to a point all was well. The turning moments and out-of-balance forces were not significantly different from those on which Husband had based his calculations on the snowed-up mesh. The wind tunnel had, however, shown up a new difficulty. In November when I talked with Scruton at N.P.L. he asked me if I had heard of the type of structural flutter which, some years previously, had destroyed the Tacoma bridge in Washington State. When I mentioned this to Husband he said that the natural frequency of the telescope structure was probably so high that no trouble would arise in this way. He had, however, often discussed with me his anxieties about the teeth on the elevation driving racks and pinions and I knew that he realized that this would be the first part of the telescope to give trouble from wind effects. Scruton's flutter tests on the model confirmed both Husband's belief about the safety of the main tower structure and, alas, his anxiety about the bowl flutter. In the model, springs were used to simulate the support of the bowl from the towers and Scruton had succeeded in destroying the springs through structural flutter in a wind speed of only 40 m.p.h.!

That was the background to the item of 'braking' which suddenly appeared in the new 1955 April estimates. Husband told me on that Sunday afternoon that he now had no alternative but to take urgent steps to introduce some form of braking or damping into the bowl movement. At that stage he proposed to use two extra pinions, which were already available, connected to braking drums carrying ton weights. It was the cost of this work, estimated at £15,300, which appeared in the estimates. The figure of £70,000 which Husband was now estimating for the driving system included a revised cost of cables and it still seemed, even at that dark moment, that provided we could avoid the import duty on the Siemens machinery and get the extra £50,000 from D.S.I.R. that the overall deficit might not amount to more than £15,000 or £20,000. Rainford was dismayed and angry and the meeting on the morrow with Husband seemed likely to be stormy.

Until the time of that meeting on 5 April at Jodrell Bank I could still persuade myself that in spite of all the cost increases and the introduction of the braking we could finish the telescope in a steerable condition for a sum of about £70,000 over the original estimates, of which we had been promised £50,000 by the D.S.I.R. At the end of that morning, even by summoning all my resources of optimism, I could have no greater ambition than to finish the telescope with the surface on and see about driving it afterwards, even if only in a crude fashion. Rainford asked Husband for an explanation of the £70,000 including the braking item. Later in the day I wrote:

> Husband said that this had been included as a result of recent wind tunnel tests at N.P.L. on the redesigned bowl, and moreover, that some of the £3000 which he had estimated as the cost of the steel work had already been used for the top of the towers. This led on to a discussion of the overall over-expenditure as detailed in Husband's letter to Rainford in November and at that stage the real trouble began. Whereas in that letter Husband had given his estimate of the overexpenditure as £15,000 he now said that he had been forced to put another 100 tons of steel into the bowl. Even although I had perhaps some idea that Husband's £15,000 was insufficient, nevertheless this news staggered me as much as it did Rainford. All the information I had was that the complete bowl design had been with the United Steel Structural Co. before Christmas. . . . Husband agreed that it would be unwise to place any further orders for the mechanical system until the situation about the excess expenditure on other parts of the structure had been sorted out.

That was a terrible morning, at the end of which Husband agreed that he would let Rainford have his final estimates of steel tonnage and all other items of over-expenditure in two or three weeks' time—this to include the contractor's estimate of the cost of the solid membrane on the bowl.

The darkest months

The appalling troubles in which we were landed had led to a situation in which Husband and I were beginning to make our separate cases to Rainford. Husband's case was quite straightforward—namely that I was responsible because of the changes I had made in the design of the telescope. My case was not simple since I had obviously continuously discussed changes with Husband, and although I could produce a vast array of memoranda and letters containing the essential story of the previous chapters, my position was a weak one. I should say here that Rainford and the University Officers never wavered in their solidarity on behalf of the telescope and myself. It was a corporate act of faith which still had to suffer its greatest strain and is probably unparalleled in the history of scientific projects.

At the end of the day on 5 April Rainford could no doubt have spoken to his financial colleagues in London, expressed his alarm, and asked for the whole project to be stopped. He did not, firstly because he too was enthusiastic about the telescope and wanted to see it finished, secondly because the measure of our debt still did not seem unmanageable—£70,000 with £50,000 promised from D.S.I.R. Before taking action on the financial front he wanted the complete picture including the membrane cost and he expected to get it in two or three weeks.

In fact it was five months, not three weeks, before the real issues were faced. There were at least two reasons for this. The first difficulty was that Rainford was quite rightly insisting that the next set of figures should include a contractor's price for the membrane. The firm of Dempsters had been introduced to this problem. We made a mock-up of a small part of the framework on which the steel sheets would have to be fixed and Dempsters carried out a sample series of welds for their own assessment and so that we could measure the conductivity across the joins.

May 5 . . . apart from the fact that Dempsters came here on Monday and removed the sample surfacing nothing further has happened on the membrane problem. . . . they have asked permission to have another shot at the welding in their works. It is obvious that Husband can't let us have any more prices until the feasibility of Dempsters' process has been assessed and until he has a price from them, hence all matters on the financial front are also in abeyance.

May 25. Goodall told me that Dempsters were hoping to attempt a second weld this week but they found that they had cut the sheets to the wrong size.

June 3. Husband said that Dempsters' new attempt at welding the membrane is still not satisfactory and they have now decided to try with a thicker sheet. Husband is convinced that they will be able to make a satisfactory job with this. He pointed out that the cost did not increase very much with thickness and that in any case the cost of raw material involved was of the order of £4000 to £5000 compared with the basic cost of £12,000 to £13,000 for the mesh.

June 30 . . . surprised to hear from Goodall that Dempsters have still not made their third attempt with a thicker sheet.

Meanwhile—

July 6. Trouble is descending on all sides again over the telescope. Lord Simon is worrying Rainford because he thinks that the expenditure is going to be so great that the University can't afford it. Rainford is writing to Husband who is now back from Ceylon but I have tried to calm him and pointed out that only 950 tons are in the substructure and that is already completed. My point is that we certainly have enough money to pay for what is being ordered, especially if we call on the extra £50,000 which the D.S.I.R. has promised.

The second difficulty was that Husband was faced with a long task, in making the detailed costings which Rainford was demanding. Everything was turning out ever more expensive—the cables for example which were of great complexity and much underestimated initially. Further, he had been forced to second thoughts about the damping mechanism. The emergency arrangements with extra pinions made after the wind-tunnel tests on structural flutter were not good enough.

June 3. It is now proposed to have a large bicycle wheel affair from the back of the telescope rim moving transversely across the diametral girder where there will be friction brakes. The great advantage of this is that it will prevent flutter in high winds but will also give us an auxiliary method of control to enable the bowl to be moved in elevation. Husband estimated that the cost of this steel work which is quite light might be about £5000 . . . general impression is that we will find ways and means of moving the telescope without the full driving system and this new arrangement in elevation is one. For the azimuth movement Husband thought that we would have to work temporarily with winches.

August 8. . . . On the train journey [referring to a meeting in London on 26 July] Rainford produced correspondence which he had with Husband during my absence. It is quite clear that a major explosion is in the offing. Rainford is demanding Husband's new estimate which he promised last April and Husband is temporising and asking that it should be made quite clear to the D.S.I.R. that the telescope has been 'redesigned' and also demanding to be present when they are presented with the new estimates. It is a most ugly situation which cannot fail to explode as soon as Husband and Rainford return from their holidays.

Also on that day:

the Dempster experiments on surfacing still continue, although Goodall told me that he thought the latest sample was likely to prove satisfactory . . . *Aug. 22.* I went to Dempsters in Stalybridge last Thursday to measure the d.c. resistance across the sample steel membrane. Husband says that this sample is correct to $\pm \frac{1}{8}$ inch and that the whole surface should be true to 1 inch when the bowl is finally assembled.

At last:

Sept. 9. This morning a letter from Husband to the Bursar's Office gave the dreadful news that Dempsters wanted £33,000 to do the membrane. Husband says that the cost of the sheet was only £5000 and that the high cost was entirely due to labour. Apparently Dempster's first figure was in the £40,000 to £50,000 region. To this must be added about £2000 for the painting. Quite clearly nothing can now delay the blow up which must occur when Rainford gets back from holiday in a week's time. Even worse Cannon said there was a certificate from Husband asking for a further payment of £56,000 to the United Steel Structural Co., making £165,000 in all for work completed. It is not clear how much of the steel for the bowl this includes, but it is already far in excess of the contract price of £131,000.

The strange aspect of all of this is that even at that stage I was still hoping to use the telescope (undriven) by the end of the year—the entry continues:

The other news in Dempster's letter was that the order would have to be placed within a few weeks if they were to get delivery of the steel sheet which would enable them to start work in April 1956 and that given good weather it would take them 6 months to complete! So much for any hopes we may have had of starting work this autumn.

The revelation of the true state of our finances was not long in coming. On Saturday 17 September Cannon phoned 'with the appalling news that a letter just in from Husband gave the weight of steelwork as 1700 tons and the cost as £221,000. I was prepared merely for 1500 tons, and this together with the news about the membrane means that the crisis is far more severe than I had ever anticipated.'

Rainford prepared a document for the Officers and Council of the University in which he concluded that another £160,000 was needed to finish the telescope.

Sept. 23 . . . one thing only keeps me reasonably calm, it is that so much of the structure is now complete that it really is inconceivable that the country won't finish it . . .

Sept. 27. Called on Rainford—who had a violent response from Lord Simon demanding an investigation of the whole issue as a matter of urgency, but he [Rainford] said that the Vice-Chancellor was away and until his return on Friday he could not take any steps.

Eventually on 7 October, Husband, Rainford, and I were summoned to Sir John Stopford's office. Sir Raymond Streat (the Treasurer) and Mansfield Cooper (who had acted as Stopford's deputy and was destined to succeed him as V-C.) were present but Lord Simon couldn't come ('. . . Rainford thought it was just as well').

Rainford had by this time increased his own estimate of our deficit to £170,000. The agitated meeting continued for ninety minutes. Husband said that the increased costs over the estimate arose because of the redesign of the telescope and that I was responsible for the redesign. Clearly I was the defendant and could only argue that the financial burden of the redesign to the best of my knowledge amounted to £50,000 for which I had subsequently obtained a promise of coverage from the D.S.I.R. Finally

. . . the V-C. asked us to prepare a document on the situation for the information of the Officers of the University only. I said that under present circumstances it was unlikely that Husband and I would reach agreement on the contents of this paper, but the V-C. said in that case there must be amendments. Streat said the situation was appalling and the only hope was to go forward with a unified picture from the University end.

The next day I wrote that the proceedings had 'left me utterly miserable'. I used to seek comfort by gazing at the great mass of the telescope. The hugeness of it, even at that time, somehow made it seem inevitable that it would have to be finished. Two years later almost to the precise day of that meeting the telescope was to be headlined all over the world because it had located the carrier rocket of the first Sputnik. Subsequently I wondered, in moments of idle speculation, what effect a foreknowledge of that event would have had in October 1955. One thing alone is certain—it could never have added more to the resolve of the University Officers to see the job finished, distressed as they were at that time.

Although Lord Simon was not at the meeting, he demanded, and got, from Rainford and Cooper a full report of the proceedings before nightfall. The Simons had already arranged to bring Sir Hugh Beaver out on Sunday afternoon. I had been urgently asked not to discuss the telescope finances with Beaver (he was Chairman of the D.S.I.R. Advisory Council). It must have been hard for me to avoid doing this but in my record of that visit I note that Lady Simon was taken straight to our home and we followed for tea two hours later. '. . . His question about money had to be side stepped. Instead I carried through my allotted task and talked hard for nearly two hours about the use of the telescope . . .' As for Lord Simon, '. . . he was far more comforting than I expected. In asides he expressed his horror of the situation but even so talked about ways and means of raising the money from industry.'

It was inconceivable that we had not reached the nadir of our troubles. But it was not so, and during those weeks it seemed that nothing could stop the escalation of our financial burden. For the Vice-Chancellor and the Officers of the University nothing was more important in the days following the meeting of 7 October than the consideration of the report for which Husband and I had been asked. Husband did not produce his part of the report with the alacrity the V.-C. anticipated. Rainford was pressing Husband daily, and on Saturday 15 October he and Cannon journeyed to Sheffield in an attempt to sort out the essential features of the report with Husband. However, the complication of the situation was great. Husband was on the eve of departure for urgent business in Ceylon and the main result of that meeting was an agreement that Cannon would go into the intimate details with Betts and Holmshaw during the following week. He did this:

Oct 21 . . . as a result of which the true situation is revealed as being more disastrous than ever. Two things make it so, firstly the United Steel Structural Co. are claiming £135 per ton instead of £114 per ton in the contract. Secondly the weight of steel has again gone up to 1660 tons, i.e. nearly 100 tons more than was being talked about when Husband came over to see the Officers. The net result according to Cannon is that we are now £240,000 in arrears, the new estimate being £630,000 compared with the £390,000 of 1953 and with the £240,000 of 1952. The situation is, of course, appalling. If orders were not placed for the driving system we should save £60,000 and on the membrane £20,000. That means £160,000 over, and with the £50,000 promised from the D.S.I.R. leaves £110,000 to be found from somewhere even to cover the basic steelwork. All the old worries about the work being stopped return, alleviated only by the thought that this would not rectify the situation since all the steel is now cut and waiting for transport to Jodrell.

This revelation of the real state of the telescope finances naturally caused the utmost gloom and despondency amongst all concerned in the University. The Vice-Chancellor would clearly have been justified in dismissing me with ignominy. It is a measure of his great stature that he did not, and in fact I doubt if such a course of action ever entered his thoughts. He wrote to me a short letter in which he merely said that the situation was 'simply terrible' and he did not know where the money would come from. In fact I doubt if those responsible for guiding the affairs of a University have ever reacted so quickly and so solidly as the University of Manchester Officers did at that moment.

Oct 27 . . . last night at the V-C.'s reception things seemed a little brighter. The Treasurer, L. P. Scott[1] and Rainford all seemed to be taking the measure of it. I am to go to lunch on Monday with the Officers, at which the University policy will presumably be decided. It is a tremendous relief that all the Officers seemed determined to see the job through, although as the Treasurer said, reputations would suffer especially in a University which had such a good reputation for its excellent handling of financial affairs.

As for Lord Simon he seemed entirely concerned, not with the magnitude of the debt, but with whether this implied a basic fault in the design. 'He was relieved when I assured him that if the financial trouble could be overcome we could use the telescope next year.'

The next day (27 October), Rainford had arranged to bring the Chief Finance Officer of the D.S.I.R. (Smith) to Jodrell. Smith gazed out of the control room window and then turned to me: 'the strength of your position, Lovell, is that huge mass of steel'. Eventually he said that the D.S.I.R. had the £50,000 in reserve for us and then gave his

[1] Laurence Scott, the Chairman and Managing Director of the Guardian and Evening News Ltd., was also deputy chairman of the University Council.

personal opinion that if the Advisory Council could be persuaded it might be possible to do a pound-for-pound offer on the remaining £200,000—implying that we would have to raise £100,000 ourselves. 'On his way back from Crewe, Rainford phoned to say that he felt the afternoon had been well worth while. He seemed greatly cheered up.' In fact this was not the first D.S.I.R. job to run into a colossal over-expenditure and the situation was not new as far as Smith was concerned. Nevertheless the problem of putting into action the procedure suggested by Smith was not easy, and as far as D.S.I.R. was concerned there had first to be a most searching inquiry.

1. DECEMBER 1945. The first day at Jodrell. The receiver trailer with aerial attached outside the wooden botany hut.

2. JUNE 1947. An early morning at the time when the summer daytime meteor streams were discovered. The 'searchlight aerial' is to the left and the trailer (with prime mover) is the 'Park Royal'. It was into this cabin that I led the Vice-Chancellor and Blackett.

3. The original 218-ft. diameter wire paraboloid at Jodrell Bank. The paraboloidal bowl was fixed to the ground but the beam direction could be changed a limited amount by altering the tilt of the mast carrying the aerial at the focus 126 ft. above the ground. Dr. C. Hazard is on the left and Prof. R. Hanbury Brown in the centre. (Copyright Central Press Photos Ltd.)

4. 1948. Jodrell Bank in the summer when we were still working in caravans and trailers. The 218-ft. transit telescope, the searchlight aerial and the radiant aerials on the borrowed strip of land (see p. 30) can be seen. Field 132 in which the telescope was eventually built is at the right; the position has been marked.

5. 25 AUGUST 1952. The peaceful scene in Field 132. That night the contractors' hut seen in the background was demolished and thrown into the ditch (see p. 40).

6. An artist's impression of the original design of the radio telescope using an open wire mesh reflector. (Copyright Husband & Co., Sheffield.)

7. 8 NOVEMBER 1952, 'the weather remains wet and windy and the whole site is a quagmire'. At work on the sixty-seventh pile. The author and his family are on the right of the picture.

8. DECEMBER 1952. The reinforcement in the central thrust block and annular chamber. The tunnel to the control room leads out of the photograph at the top right.

9. The reinforcement for the concrete circle on top of the piles which was to support the railway track. On 31 July 1953 the last gap in the 350-ft. circle of concrete was closed.

10. The laying of the double railway track began in mid-January 1954 and was completed by mid-March.

11. On 11 May 1954 the central pivot was delivered to the site. Here it is seen being fixed in position on the central thrust block.

12. MARCH–APRIL 1955. One of the last undriven wind carriage bogies to be delivered ready to be rolled under the wind rakers of the northern tower.

LEFT

13. 10 AUGUST 1954. The cranes have just placed the large girder assembly on the central pivot—the first of the steelwork to be erected.

BELOW

14. EARLY OCTOBER 1954, the chord girder on the southern side was completed and the cranes began lifting the tower sections into position.

15. MARCH 1955. The crane is still attached to the main trunnion-bearing girder which was lifted to the top of the southern tower on March 15.

16. 25 APRIL 1955, when the construction of both towers had been completed to the trunnion-bearing level. The stone tracks on which the cranes moved are clearly seen, and also the shell of the control building. (Aerofilms Ltd. Copyright reserved.)

17. JUNE 1955. Cooper's engineers assembling the trunnion bearings.

18. AUGUST 1955, as the mass of scaffolding began to grow. Here is the tower built over the central pivot and diametral girder to carry the hub of the bowl which can be seen, waiting to be lifted, on the top of the diametral girder.

19. 14 SEPTEMBER 1955. The complete assembly of the trunnion in the bearing (Plate 17) at the beginning of its lift by two cranes to the top of the northern tower.

20. 14 SEPTEMBER 1955. The trunnion assembly of Plate 19 about to be lowered into the pedestal bearing at the top of the northern tower.

21. 14 SEPTEMBER 1955. The two cranes have just hoisted the trunnion bearing to the northern tower. The gun rack can be seen on the ground near the base of the right-hand crane. It was hoisted on 30 September.

22. AUTUMN 1955. On September 14 the trunnion assembly was lifted to the top of the northern tower, the great gun rack followed on 30 September and on 2 October the 30-ton girder assembly—the meeting point of the rack and the bowl. The central hub was lifted to the top of the scaffolding tower on 26 October.

23. MARCH 1956. The central pair of ribs of the bowl framework in position, fixed to the central hub of the bowl held on the scaffolding tower rising above the diametral base girder.

24. THE END OF MAY 1956. The second of the secondary NE ribs (No. 7 in diagram on p. 139) was placed in position on 27 May. The circumferential connecting panel (joining ribs 5 and 7 in diagram on p. 139) on which the men are working was hoisted on 24 May. On the NW side, one secondary rib (No. 6, hoisted 16 May) and the connecting panel to No. 8 rib (not yet in position), hoisted 27 May, can be seen. The purlins on which the membrane was to be fixed were already in position on this northern section between ribs 5 and 6.

25. JUNE 1956. On 21 June rib No. 10 (p. 139) was hoisted and the main framework for the northern section of the bowl was complete. At this stage the EW line was straddled, but in the southern section only the main central ribs were in place.

26. EARLY SEPTEMBER 1956. The moment when all the major members were in position and the bowl was ringed with steel. It was supported on 16 scaffolding towers under each rib containing 90 miles of scaffold tubing. Shortly after this photograph was taken the dismantling of the large cranes commenced. (Reproduced by kind permission of *The Guardian*.)

27. The $62\frac{1}{2}$-ft. aerial tower, which is seen here surrounded by scaffolding, had to be lifted while the cranes could still command the central area of the bowl. It was placed in position in mid-July 1956 before any of the ribs in the southern section of the bowl had been lifted. Orthostyle left the site at Christmas 1956 with only 80 ft. diameter of the membrane in position as shown here. They did not resume work until the spring of 1957.

RIGHT
28. MAY 1957. The swinging laboratory is hoisted into place underneath the bowl.

BELOW
29. MAY 1957. Welders at work on the membrane.

30. LATE IN MAY 1957, Messrs. Orthostyle at last began to make rapid progress with the fitting of the membrane. Here on 31 May the welders can be seen at work in the bowl with approximately 17 more rows of plates to be fixed. The large cranes have disappeared and the bowl is still supported on two scaffolding towers. (Aerofilms Ltd. Copyright reserved.)

RIGHT

31. 1957. The scene inside the bowl during the summer. Some of the steel plates for the last 9 rows of membrane are lying in the bowl near the welding equipment. Our first aerial can be seen on the top of the central mast at the focus.

BELOW

32. 20 JUNE 1957. The bowl of the telescope is moved in elevation for the first time. The membrane is incomplete and some of the supporting scaffolding is still visible.

33. 1957. The scene in the control room during the early summer when the indicating racks and console were being wired. One half of the indicating racks (actual and demanded azimuth and elevation, and universal time) is visible and an edge of the control console on the right.

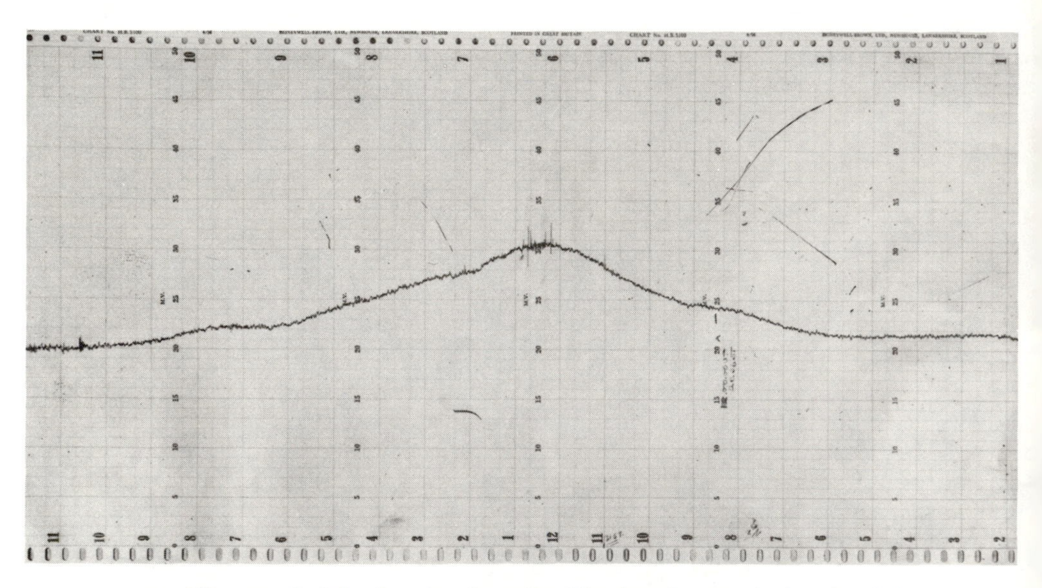

34. The record of the first signals received by the telescope during the night of 2 August 1957. The bowl was fixed in the zenith and the peak in the record occurred as the movement of the earth swept the beam of the telescope across the plane of the Milky Way.

35. REFLECTIONS OF A ROCKET. The radar echo from the carrier rocket of Sputnik I located by the radio telescope on 12 October 1957. The author is seen explaining the record to the Press Conference at Jodrell Bank on 13 October 1957 (see p. 198). (Reproduced by courtesy of *The Guardian*.)

36. THE MOMENT OF IMPACT. The telescope at 10 p.m. B.S.T. on the historic Sunday evening at Jodrell Bank 13 September 1959 as the Soviet rocket Lunik II hit the moon. (Reproduced by courtesy of the *Daily Mail*.)

37. Several Russian scientists visited Jodrell Bank during the construction of the telescope. Here the distinguished Russian astrophysicist, Professor Alla Massevitch, is shown with the author at Jodrell Bank on 19 February 1960. At that time Professor Massevitch was in charge of the satellite optical tracking network in the Soviet Union. (Copyright Press Association.)

38. 18 MARCH 1960. Her Royal Highness Princess Margaret presses the button which initiated the command signals from the telescope to Pioneer V, 1,400,000 miles away in the solar system. Mr. William Young, the head of the American team working with us at Jodrell Bank, is timing the operation. (Copyright *Daily Mail* Manchester.)

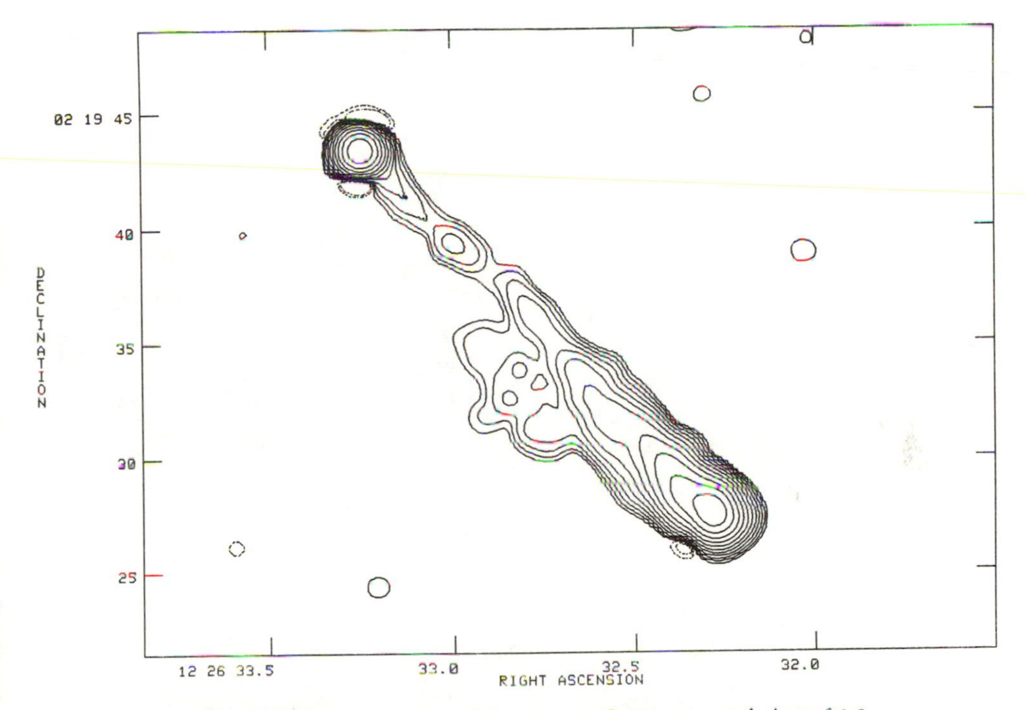

39. The MERLIN contour map of the quasar 3C 273 at a resolution of 1.0 arc-second. The jet extends from the core for more than 100,000 light years. There is not yet any agreed explanation of the physical processes by which the jet is produced or of the processes of energy production in the core and jet. The core of 3C 273 is not resolved in this map and subsequent observations with one thousand times the resolving power have shown that there are still unresolved components of angular extent less than a few milliarcseconds. (This contour map is reproduced from the paper on MERLIN by P. Thomasson in Quarterly Journal of the Royal Astronomical Society, 27, 413, 1986 by kind permission of the author and Editors. The original research was by R. J. Davis, T. W. B. Muxlow and R. G. Conway, *Nature* 318, 343, 1985. The contours in the above (reading from the outside) are at intensity levels of (1, 2, 4, 8, 16, 32, 64, 128, 512, 1024) x 0.05% of the peak intensity of the unresolved core which is 13.4 Jy per beamwidth (1 Jy (Jansky) = 10^{-26} watts per sq. metre per hertz). The observing frequency was 408 megahertz.)

40(a). The original Mark I telescope photographed in June 1961. When the telescope first came into use in the autumn of 1957 a tubular steel mast, sliding into the aerial tower structure carried the dipoles at the focus. With this arrangement (which can be seen in Figure 31) about 80 ft. of cable was required to connect the dipoles to the electronic equipment in the elevated laboratory underneath the bowl. To achieve the sensitivity necessary to command and track the American space probe Pioneer V (Chapter 35) it was essential to avoid the losses in this cable and the tubular mast was replaced by a framework to hold a box containing the electronic equipment with the aerial mounted underneath at the focal point as shown in this photograph. The control room is at the bottom left of this photograph. The trailers and temporary hut containing the American equipment can be seen to the right of the control room. The photograph in Figure 38 was taken inside these trailers. (Reproduced by permission of the Controller of Her Britannic Majesty's Stationery Office.)

40(b). The Mark IA telescope. The major changes are the new reflector mounted above the original which is still in place. The aerial tower has been lengthened and strengthened to carry a heavier load of equipment near the focus. The slender stabilising circular girder (the 'bicycle wheel') seen in Figure 40(a) has been replaced by the massive double load-bearing wheel girders. The diagonal braces from the towers to the diametral girder which were introduced after the hurricane of 2 January 1976 are clearly visible.

22

1955 on the site—a beginning with the bowl

THE unfolding of the financial chaos in 1955 occurred against the backcloth of the growing telescope steelwork. Indeed it seemed as impervious to the financial troubles as it was to the winter winds. I often wished that our meetings could be under its shadow. The growth of our deficit seemed less important when measured against the actuality of what had already been achieved.

The wind rakers for the northern tower were being erected in January, otherwise, apart from the serious problem of the riveting, the steelwork seemed fairly complete to the 140-ft. level. A lot of steelwork was still being delivered but the delivery promises for the top 40-ft. sections of the towers had been broken. This material was promised before the end of 1954 but by mid January of 1955 Goodall was expressing doubts as to whether the steel for these sections would be on site even by the end of the month. Since the target was to complete the towers entirely by that time, the programme began to fall behind schedule. Husband was more concerned about the riveters who gave endless trouble because of their autocratic behaviour and endless strikes; indeed he said then that he thought the riveting situation might well govern the overall progress with the telescope.

Jan. 20 . . . situation on the erection somewhat gloomy. Although some sections for the final parts of the tower have arrived news came this week that the erection people have been ordered to work only a 44 hour week. This is the employers' reply to the sporadic strikes which have been going on throughout the country. The gravest aspect of this matter is that some members of the team say that it is not worth their while working here for only 44 hours and intend to return to London . . . *Jan. 24* . . . situation over the work of the steel erectors is simply agonising. For almost the first time this year the weather during the last few days has been very good and yet the steelworkers finished at 12 on Saturday and 4.10 this afternoon; thus instead of having the entire north western wind raker erected they are still only halfway up. Husband is also very depressed about the riveting and estimates that so far only about 10% of the total amount of riveting required in the lower part of the structure has been carried out.

In actual fact a few fine days made a big difference. Before the end of the month the wind raker and the top of the northern tower were complete. Sometimes it was quite impossible to work on the structure. '*Feb 21*. With Husband on the site in two inches of snow and an icy wind. This week there has been absolute stagnation; not a single bolt or rivet has been driven and the steel erectors have been huddled in their cabin.' Husband remained optimistic, 'he still feels that in spite of the current delays on account of the weather and the difficulties about the changes to the top sections consequent upon the introduction of the brakes that the United Steel Structural Co. must in their own interest erect the bowl this summer.' The Managing Director of the United Steel Structural Co. wrote to me at that time and said that the bowl was now in their shops and that they were hoping to erect it during the summer. These estimates and hopes were wrong, but I'm glad that I did not realize that during those months.

On a spring day in early March when rivets were going in at the rate of 1,200 a day and men were again working on the tops of the towers I made an estimate with Goodall of the time needed to complete the rivets in the base and tower sections. Then about two-thirds of the diametral girder had been done representing a third of the total: 12,000 done, another 36,000 needed, requiring twenty-four days at that rate of progress. On 10 March: '. . . a significant occasion when the racks and the supporting framework from the battleships, which have been in our workshop for so many years, were loaded for transport to Scunthorpe. Goodall thinks that these may be back again in about ten weeks time.'

March was a good month. On the 15th a 12-ton girder containing the holes for the two pinion axles was hoisted into position on the south tower. By the 22nd I noted that this tower was complete apart from the last heavy girder and all the time large amounts of minor strengthening steel, including the access ladders, were going into position. The main trunnion-bearing girder was in place before the end of the month and attention was transferred to the northern tower. This, too, reached the trunnion-bearing level a few days before April began to unfold the depressing financial story.

The structure had now reached the stage where it seemed likely to remain virtually the same for some time. The riveting and levelling had now to be completed and the next stage in which the driving racks and cones, bearings, and trunnions had to be fitted was realized to be one of the most difficult and critical of the whole erection process. During the whole of April riveting was the main occupation on the structure. The steel erectors had little to do except for the installation of minor parts like handrailing and walkways and some were paid off. Indeed, even with optimistic forecasts it seemed unlikely that much more visible progress could occur until June. It was unfortunate that this pause in construction occurred during the good months of the year.

It was during this time that the ancient oak tree outside my office window achieved fame. When the levels were checked it was found that the southern tower was 2 in. out of vertical at the top. It did not yield easily to struggles with the hawsers until one wise man decided that the oak could take the strain. It did, and today shows no visible sign of the huge wire rope that encircled its trunk on those April days.

June 2 . . . Whit week and the railway strike are not very good combinations for the radio telescope. Actually the riveters have almost completed their work and it is extremely sad that there should be a shortage of material at this stage . . . [The trunnion bearings, housings and racks were overdue.]

June 13 . . . Husband wrote that the effect of the railway strike was beginning to make itself felt rather badly . . . Robeys who have agreed to do the machining on the big racks have had to reorganise the whole of their work schedule because of the necessity to stop overtime on account of the strike. They have informed Scunthorpe that the machining of the rack which they intended to do in June would now have to be delayed until the end of July.

In fact a compromise was soon reached by which Robeys agreed to do the machining in the first week of July before their works holiday after which the second rack should have been prepared for machining by Scunthorpe.

In mid June the steelwork for the central hub of the bowl arrived and since this was to be lifted in one piece the assembly of the central 20-ft. ring of the bowl was immediately laid out on the ground. Apart from the background anxieties it would have been an exciting stage since the main trunnion-bearing assemblies were also being delivered at this time and there still seemed a chance that the bearings and rack assemblies could be in place to make a beginning with the main bowl framework by the end of August.

June 30 . . . the actual situation on the telescope construction changes extremely slowly. Last week Coopers fitted the part of the bearings which clamps on to the trunnions, and to everybody's surprise, except apparently Cooper's, these fitted perfectly and the engineers left them covered up on the site. They have now gone away to do the final machining on the rollers. There has been a hold up in the delivery of the pedestal bearings and it is now not expected that these will be hoisted in position until the steelworkers return from their holiday at the end of July.

Another month of holidays and delayed deliveries led to a stagnant period—a source of great irritation that advantage could not be taken of so much good weather. In early August the pedestals were on site, one in place at the top of the southern tower, the other on the ground. The racks had still not arrived from Robeys and the bowl erection was receding into the mists of winter.

August 22 . . . the hold up on the erection is serious. It is entirely due to the delay in the supply of the rather small bearings required for the trunnion bearings . . . this delay has scotched all hopes of doing an early test on the telescope in its static condition. Husband is reluctant to commit himself but it seems clear that there is no hope whatsoever until late next summer.

The hoisting of the heavy trunnions, the racks, and the associated steelwork required two cranes working in synchronism and after many delays because there was only one crane driver the first trunnion assembly was hoisted into position on the northern tower during the morning of 14 September. On the last day of September the great gun rack was lifted and two days later the 30-ton girder assembly—the meeting point of the bowl and the rack. The days were calm, and the sight of these massive objects imperceptibly moving upwards against the clear blue sky remains one of the most vivid memories of the whole construction.

During October, as our true financial state was being revealed, a large amount of steelwork for the bowl was delivered and the first rib was laid out on the ground and partially riveted. The process of erection had of course been decided. In principle the central 20-ft. circular hub of the bowl was to be lifted in place and supported on a tower of scaffolding built up from the ground and the diametral girder of the telescope. From this the radial ribs of the bowl would be mounted with their peripheral extremities supported on the huge mass of circular scaffolding which was soon to encase the telescope structure.

On 26 October the central hub, which had been built up and riveted at ground level, was lifted by the two cranes to the top of the central scaffolding tower. In preparation for this the scaffolders had been summoned in early August.

Nov 8 . . . the processes on the site have gone through a very trying time recently since most of last week was spent in trying to centre the rack to $\frac{1}{8}$ inch. This was achieved and the rack bolted into position by the weekend. There is some hope that the first two ribs from the centre hub to this end piece will be raised this week.

A month later we were still hoping that the ribs would be raised 'next week'. Alas the calm weather of autumn had vanished and the opportunity to carry out heavy lifts and work at high level occurred rarely.

Moreover there was still continuous trouble with the riveters. There were even cases when a gang would travel hundreds of miles to Jodrell Bank, take one look at the place, and leave within the hour. Their excuses varied. Either it was too far from a town or they could earn better money 'riveting a cowshed in comfort'.

On 19 November the first rib was lifted to a vertical position and tied against the cross girder in preparation for lifting.

> . . . What a sight! The partially finished second rib is now over on the East side and the assembly is nearly complete. If the riveters come back there would seem to be a chance that the ribs might be lifted next week. . . . *Dec 2* . . . the ribs still not hoisted. Meanwhile the United Steel Structural Co. have decided to send all the bowl steel here; a new mobile crane has come and is unloading tons every day. The station at Holmes Chapel has had to import a special crane!

At last on the calm and sunny Sunday morning of 4 December the first rib was hoisted and the second one four days later. All went without a hitch. The next main task was to hoist the two equivalent ribs on the south side so that the central hub would be securely fastened to the rack girder assemblies, but this implied the lengthy job of shifting the cranes across the site, and while they were in the northern position it was decided to use them to do infilling of the smaller steel members which braced across between the ribs, and also to lift the steel of the walkways and other miscellaneous items. The year ended in this way, a far cry from the expectation that the telescope would at least be usable in a stationary condition. 'Christmas Eve 1955—another Christmas without the telescope—but surely not next year! The site is littered with hundreds of tons of steel for the bowl—most of it must now be on the site, but only two of the ribs are up.'

23

Appeal and inquiry

RAINFORD had brought the D.S.I.R. Finance Officer to Jodrell on 27 October, the day after the central hub of the telescope bowl had been hoisted to its supporting tower. The anxiety over our finance continued unabated as the telescope structure proceeded. On 31 October I was summoned to luncheon with the University Officers.

> Lord Simon was in a fairly fierce mood and immediately tabled a list of 30 questions about moral and legal responsibilities for the present situation . . . right at the end I was instructed to explore ways and means of raising the money from industry etc. and was then dismissed. In the quadrangle afterwards I met Rainford who said not to worry since he had seen Simon write £10,000 in his note book, an action which Rainford took as indicating his intention to assist.

It seemed to me that my instructions were clear, namely that I had to set about raising a sum which, at that moment, we believed to be £100,000 on the assumption that D.S.I.R. would find an equivalent amount in addition to the £50,000 previously arranged. I set about the task without delay. I wrote many letters to those of my friends who were kings of industry, I had meetings with every Ministry whom I had ever assisted, and sought avenues with the Royal Society for transferring some of my International Geophysical Year allocations to the telescope on the grounds that the telescope was vital to the I.G.Y. programmes.

Although Rainford had already written to Cawley in D.S.I.R. to explain our intentions and Blackett would in due course have learnt of the situation officially since he was a member of the Advisory Council, it was by accident through the I.G.Y. avenue that he first heard what was happening. My initial probes into the I.G.Y. issue had left me feeling reasonably hopeful that I might manage to get support in the £10,000 to £20,000 region. After several interviews with Sir David Brunt, who, as the Physical Secretary of the Royal Society, had arranged the Treasury support for the Royal Society I.G.Y. programme, he suggested that I should put the I.G.Y. story to the D.S.I.R. with a copy to Blackett. The urgency was such that I sent these documents by special delivery arrangement on 14 November (a Monday). '*Wednesday Nov. 16* . . . this morning I had a tremendous blast from Blackett over the phone who said he had never known such a mess and incompetence . . . he accused me of not spending enough time on it (!) and of trying to do other things as well. He also returned to his old hobby horse that I had insisted on making it too big against his advice. Not one of life's happier mornings.'

An unhappy Sunday afternoon

In Chapter 19 I mentioned that on one occasion my apprehension which accompanied Lord Simon's visits turned into a frustrated terror. The occasion was a winter's Sunday afternoon—13 November 1955. Lord Simon was a man of instant action. He did not leave me in the slightest doubt about his opinion of the incredible situation into which I had landed the University. But he was fascinated by the telescope project and once he was assured that the troubles really were only financial and not scientific or engineering ones he was determined that it should be finished. At the beginning of this chapter I mentioned that at the luncheon party on 31 October after appearing at his fiercest he apparently took out his notebook after I had left and wrote £10,000 in it. His next move was to assemble twelve directors of Henry Simon Ltd. at Jodrell Bank. I imagined that it must be quite difficult to collect together the directors of a large firm at short notice. In any case the Sunday week following the luncheon I met them at 2 p.m. precisely in a freezing wind outside the power house at Jodrell Bank. I had been instructed that my duty that afternoon was to impress them with the necessity of starting the telescope fund with a generous donation of £10,000.

It did occur to me that perhaps from their point of view their own firesides might have seemed a more attractive place than a bleak and bankrupt Jodrell on that Sunday. However, I suspect that some at least had the same kind of healthy respect for Lord Simon as I had myself. The proceedings began badly and ended disastrously. Lord Simon was late! The directors shivered in consternation. He had never in anyone's recollection been more than a minute late without communicating the fact. Alas, Jodrell has two entrances. I had met the directors at one entrance and Lord Simon's chauffeur had driven him to the other.

It was a terrible start and the assembly must have found it as hard to concentrate on the problems of the universe as I did to speak to them in that frozen state. However, the worst part of the afternoon lay ahead. I had planned to walk from the control room to the site of the telescope which at that moment had reached the tower top level. In order to impress the directors with the work already accomplished my return route was to be down the steps underneath the central pivot and along the cable tunnel into the basement of the control building. This basement eventually leads back into the entrance foyer of the control building where my wife was to be waiting with tea.

Unfortunately for me I had issued standing orders that the engineer on duty was to check that the door at the top of these steps was always locked and that the door at the other end of the tunnel was also locked at weekends, in order to prevent unauthorized people entering into the building. Before leaving the building with the party I had of course checked that the doors were open. Alas I had not anticipated that the efficient duty engineer would in mid afternoon check these doors. He did just that. First, seeing no one around, he locked the door at the control room end. Then with me already in the tunnel with Lord Simon and his party he proceeded to the central pivot and locked that door.

I reached the dark steps at the end of the afternoon's journey. I said that if Lord Simon would please permit me to lead the way we would again enter the relative warmth of the control building where tea would be served. The door was locked—I was stranded with Lord Simon and twelve directors on the steep, dark, and cold steps leading out of the tunnel. I apologized and said that if they would stay there for a few minutes I would return along the tunnel and re-enter the control building over ground. The tunnel is 200 yards long. It was dark and rather wet, and when I reached the other end that door too was locked. We were all securely locked in the tunnel. Nowadays it would be easy to get out. Then the place was deserted, no telephones and no one on duty apart from the shift engineer who had dutifully returned to the power house. I returned to the party and hammered and hammered on the exit door hoping that my wife, who was the only other person in the building waiting with the tea, would eventually notice that we were not only late, but had vanished. She did—and we were rescued.

Lord Simon already had little opinion of my ability as an organizer and then I reached the depths. There is a golden end to that story. The next morning Lord Simon phoned to say that all had gone well at his directors' meeting and that they had agreed to vote £10,000 to the telescope fund.

Another reprimand

It seemed clear to me that at the luncheon party in the University on 31 October I had been instructed to start on the job of raising £100,000 at least, and I had taken immediate action, although not with the rapid success of Lord Simon. Every meeting, every party, all my friends became the target. I accepted and explored every piece of advice. I had recently spoken at an Institute of Directors dinner.

M told me to write to his father and everyone else on the dinner list. On Saturday I went through this with H and now have a further dozen local people to plough through. I am negotiating with R to get a talk with the Chairman of the Trust. On Saturday evening M said I ought to get hold of his deputy Chairman. I have written to C to see if he can help get him here. [Dec 5]

The local people were my undoing. '*Dec 10* . . . Trouble again. Frantic phone calls from Rainford yesterday when I was in London, but he didn't speak to me until this morning. One of the men I wrote to . . . immediately phoned the V-C. because he's also mixed up in the appeal for hostels.' Once more I was summoned early on a Monday morning to the V-C.'s office and

received a very severe reprimand. I asked the V-C. what he expected me to do after the lunch party and the answer seemed to be that the approaches were to be very, very preliminary with action depending on the result of the meeting which, at that time, we thought was to be between the D.S.I.R. and ourselves . . . I defended myself with reference to Lord Simon's instantaneous action . . . the situation is quite impossible. I'm in the vice-like grip of officialdom with the telescope in danger of being stopped for lack of a few thousand pounds next summer.

There were wheels within wheels. I had, in innocence, trodden on many corns by moving too quickly. These local entanglements were explained to me in confidence in calmer times. On that December day I saw no justice in the affairs of man and I remained sore and mystified. '[Jan 4 1956] . . . at a New Year's party I told Raymond Streat about my trouble with the V-C. He said it was certainly his understanding of the lunch party that I had been instructed to get the money.'

The inquiry

We had hoped that, after the D.S.I.R.'s finance officer's visit to Jodrell Bank on 27 October, his suggestion of a pound-for-pound contribution from D.S.I.R. would be put into effect without further trouble —or at least with no more than additional local internal discussions between the University and D.S.I.R. Their Advisory Council met in mid November. On 22 November Rainford and I had letters from Cawley who assured us that we had his sympathy; 'nevertheless the present position raises a number of difficult financial questions which are now being considered at a higher level than mine.'

These considerations had one immediate result. Before D.S.I.R. would even consider further grants there had to be a Committee of Inquiry with an independent Chairman. The depressing news came to me from Rainford on 5 December and his speculation as to whom the Chairman might be gave me no comfort.

Dec 6 . . . full of depression and anxiety about the proposed Committee of inquiry—one might as well have embezzled the money or be a pedlar . . . Rainford insisted on taking a reasonable view of the situation and thought there was nothing else which could be done . . . he horrified me by saying that they would probably want to go through the files. . . . a very grim and grey situation . . . the evening brightened by a letter from Blackett which I expected to be a blast but which was actually a very understanding one, giving his personal view that the Government would eventually fork out the money, but suggesting that the University might take a chance and place the outstanding contracts, if necessary borrowing the money to do so.

The inquiry was fixed for 18, 19, 20 January. On the first day the technical assessors to the Committee were to visit Jodrell and the remaining proceedings were to be in the University. In fact, some of the technical assessors made an earlier start on their business. On 3 January 1956 Thomas[1] from the Building Research Station and Scruton of N.P.L. were at Jodrell with Husband, and Saxton[2] of the Radio Research Station came on the 5th and 6th of January. At least to the outside world Husband remained completely unruffled by the proposed inquiry. I looked on it as an approaching hell on earth which could not fail to have disastrous results both to the telescope and to those of us involved in it.

A few days before the inquiry the constitution of the Committee assumed a more friendly aspect. The independent chairman and the external structural engineer assessor both withdrew (for reasons which remained unknown to me). The D.S.I.R. had to seek for replacements urgently. Finally Mansfield Cooper (our future Vice-Chancellor) agreed to be Chairman and Prof. J. A. L. Matheson, the University's Professor of Engineering, took the place of the engineer. When this had been decided Matheson telephoned to inform me himself. He was apologetic and seemed to expect me to resent that a colleague on Senate should be on the inquiry committee. I was delighted and both then and in the following years he was a tremendous help.[3]

[1] Dr. F. G. Thomas, M.I.C.E., who was then in charge of structural engineering research at the Building Research Station, Garston, Herts. He is now Deputy Director.

[2] Dr. J. A. Saxton left the R.R.S. in 1964 on his appointment as Director of the U.K. Scientific Mission in Washington but returned as Director of the Station (then renamed Radio and Space Research Station) in 1966.

[3] In 1959 Matheson left the University to become the Vice-Chancellor of Monash University in Australia.

The committee which finally assembled at Jodrell Bank on the morning of Wednesday 18 January 1956, consisted of Mansfield Cooper as Chairman, Matheson, Dr. R. L. Smith-Rose Director of the Radio Research Station, Verry[1] from D.S.I.R. (whom I think also represented the interests of the Nuffield Foundation), P. D. Greenall of D.S.I.R. as Secretary, Smith the Chief Finance Officer of the D.S.I.R., Saxton (R.R.S.), Scruton (N.P.L.), and Thomas (B.R.S.). After a brief inspection of the telescope, control building, and power house (the only items involved in the telescope finances) the committee returned to the University and began its deliberations in the afternoon.

Apart from the periods when I was summoned to the Committee room I know little of these deliberations. At the beginning of the meetings on Wednesday afternoon I was asked to be present at 2.45. I was shown to a waiting room and remained there alone until 7.30 when I was asked to return on the morrow. I was not used to having the affairs of the telescope discussed without my presence and I was profoundly irritated by this treatment. The next day, 19 January, my turn came and I was investigated for eight and a half hours, alone in the morning and with Husband and Rainford in the afternoon. My note in the diary about this long session is dated 20 January.

Everything was discussed and not a single stone was left unturned. Husband surprised me by his frankness, for example he conceded that the Defence intervention had not been directly responsible for the change in design but that it had stimulated the re-thinking which had led to the design being altered. He also said that at the time when he gave me the November 1954 estimates he did not know that the steel tonnage had gone up. He said it was the Spring of 1955 before he became seriously worried about this. There was a big discussion on the mesh versus membrane which towards the end turned into a discussion of how to secure the steel billets which were in immediate danger of being sold and would lose us another year. Smith-Rose seemed to be a tower of strength on our side and continually repeated the urgency of finishing the telescope in a steerable condition for the International Geophysical Year. Before the inquiry Mansfield Cooper had received a letter from D.S.I.R. emphasising that they had not lost faith in the timeliness and promise of the telescope. At the end of it all I was exhausted but nevertheless felt reasonably satisfied with the attitude of the Committee. Today they are meeting to produce their report.

Afterwards Rainford told me that his general feeling was that D.S.I.R. would somehow find half of the balance required on condition that we raised the remainder. He also felt that they would do something immediate over the membrane problem.

[1] H. L. Verry, Assistant Secretary in the Overseas Liaison Division of D.S.I.R.

Aftermath

After the tension of the Committee I suppose it was natural to anticipate some dynamic result of one kind or another. There were, indeed, to be consequences enough in due time, but in the months following those January meetings we were left to speculate. There was, in fact, one immediate consequential result of significance. My diary note quoted earlier mentioned the long discussion during the inquiry about the membrane, during which it became clear that the steel situation was such that if we did not place orders immediately we should probably lose an entire year on the telescope. Rainford was summoned to lunch with the Committee on their final day (20 January) and asked to see if the University could take action on this. Within hours he had persuaded the University Estimates and Finance Committee to allow the contract for the membrane to be placed with Messrs. Orthostyle,[1] with a cover that the University's liability should be limited to the cost of the steel (about £5,000) until such time as instructions were given for the fixing (estimated to be in June). Thus, even if the result of the inquiry was that the whole job would stop (which nobody believed), the loss would be small. I was enormously encouraged by this action. If we had lost the option on the billets the further delay to the completion of the telescope even in an undriven state could scarcely be contemplated. Now, in principle at least, every major telescope contract had been placed except for the driving system.

A few days after the meeting, on the advice of Greenall, Rainford wrote officially to D.S.I.R. asking for an extra grant of £250,000 in order to complete the telescope. We all realized that this could be little more than a formality and our hopes never rose above a pound-for-pound agreement; although at that moment apart from my arrangement made in 1954 for the extra £50,000, the D.S.I.R. had no official request for the money.

Jan. 30 . . . Blackett was here on Friday morning . . . beginning to take a more realistic view of our financial trouble. The news he gives about the money available for financing independent research is appalling. He, Beaver and Ashby[2] are to see the Lord President soon about the finance of

[1] In Chapter 21 the story of Dempsters' experiments on the solid membrane and their contract price is related. Fortunately Husband managed subsequently to obtain a more realistic price from Orthostyle which was within the allowance for this item. He reported this good piece of news at the meeting with the Vice-Chancellor on 7 October 1955.

[2] Sir Eric Ashby, F.R.S., had been the Professor of Botany in the University responsible for the Jodrell Bank grounds from 1946 to 1950. He was President and Vice-Chancellor of Queen's University Belfast at this time. In 1959 he became Master of Clare College, Cambridge.

the nuclear programme. He agreed that the telescope should be dealt with at the same time if the Committee's report was available. He thought that it was very unlikely that the D.S.I.R. would go more than half way to meet the deficit.

However, there was no sign of the Committee's report.

Feb 11 . . . In London yesterday and met Greenall walking along Piccadilly. It's obvious that very little progress has been made with the report. In fact he said that he had now got 3 days officially free from other business in order to write it and that he had been working on it that morning. . . . thought there might be a special meeting of the D.S.I.R. Scientific Grants Committee called in March and thought that the best which would happen would be for D.S.I.R. to find half of the deficit. . . . Rainford alarmed at the delay particularly as Mansfield Cooper is soon going to America for several months.

Finally, after much rewriting of the initial draft, the document was signed on 23 February and we gathered that it was to be considered by a special meeting of the D.S.I.R. Committee on 15 March. Rainford was certain that it would end by being a 50/50 offer 'his main worry was that they might exclude the £50,000 promised to me by Lockspeiser and thereby make it pound for pound for the whole excess'.

Another month of waiting and then: '*March 21* . . . Rainford said that unofficially he knew that at the London meetings last week it was proposed that D.S.I.R. should contribute half the money on condition that the University would find the other half. This is still unofficial because Treasury approval has to be obtained.'

Any expectation that the issue was now closed apart from a formality was a grievous misreading of affairs. The machinations which then followed led to an incredible situation which almost wrecked the project and nearly landed me in prison.

24

1956 and the Site Committee

THE entanglements which resulted from the Treasury delays and objections were fortunately slow in developing and belong to a later part of the story. Meanwhile the building of the telescope continued despite the fact that the up-to-date costs assembled for the inquiry showed that our deficit had even exceeded the quarter-million-pound figure which had been arrived at in the autumn of 1955. The sheer momentum and massiveness of the project saved it from stoppage and disruption.

Although the financial consequences of the D.S.I.R. meeting on 15 March awaited Treasury approval, instant action was taken on other matters. The Committee of Inquiry had expressed concern that the University had not followed its normal practice of appointing a Site Committee to oversee the construction of the telescope as it normally did for its new buildings. The details of the University reaction to this, following the acceptance of the report of the inquiry by D.S.I.R., came to me from Lord Simon on Easter Monday. Although Lord Simon frequently retired to his cottage at Hellsgarth in the Lake District for short holidays he never seemed to stop working. In any case it seemed that Easter Sunday must have been a heavy day for his secretary since on the Monday morning while I was idly watching some newly hatched chicks in the sun at home a special messenger brought to me a thick packet of papers.

The package contained a letter from Lord Simon in which he said that it was now up to me to raise the University's share of the money which by that time meant raising £130,000, and that he enclosed the copies of his correspondence with Sir Hugh Beaver. This gave the news that the University Council had appointed a Site Committee 'to make sure that the job is finished as quickly and cheaply as possible in time for the International Geophysical Year'. Sir Charles Renold[1] had agreed to be Chairman of the committee, and the other members were L. P. Scott, Deputy Chairman of Council (the managing director of the *Manchester Guardian*), Matheson, Rainford, and myself. This committee

[1] Sir Charles Renold, the Chairman of Renold Chains Ltd. and a member of the University Council. He died in September 1967 at the age of 83.

was to have many frantic meetings, a lot of work, and little pleasure from its duties. The burden was already on Rainford and me, but I remain full of admiration for the manner in which the other three so willingly and voluntarily gave their time and energy to the interests of completing the construction of the telescope with the maximum alacrity.

April 9 . . . meeting with Lord Simon, Raymond Streat, R. B. Barclay (deputy Treasurer), Sir Charles Renold and the Bursar. Renold's terms of reference were emphasised as being to get on with the job within the scope of the revised estimate of £650,000. There is one small snag—the formal Treasury approval has not yet been obtained for the D.S.I.R. £130,000 contribution but Rainford isn't worried about this and expects it any day. So everyone accepts that we've now got to raise £130,000—in other words back where we were in the autumn. Lord S left it to us.

The completion of the bowl framework

Renold came to Jodrell on 12 April to have a preliminary look at the shape of things before the first meeting of the Site Committee on the 20th. Already much had changed since the beginning of the year. In early January the cranes were shifted over to the southern part of the telescope. The trunnion bearings, rack, and their associated pieces had been lifted to the northern tower in September and October. The shifting of the cranes on their own tracks was a time-consuming job and so the completion of the top of the southern tower had been left while the cranes worked on the northern ribs.

On 11 January the trunnion bearing was lifted to the southern tower. The battleship rack was assembled but not riveted on the ground and the large end girder panel was in process of assembly. Then we entered a period of absolute pantomime with the riveters. Three weeks later the rack and end panel were still on the ground.

Jan 30 . . . disastrous standstill owing to the failure to obtain any riveters. Both the rack and panel for the southern tower are ready on the ground but the steel workers have no room to do anything else until these are out of the way . . . *Feb 2* . . . Yesterday morning the happy sound of riveting emerged from the site . . . *Feb 3* alas the happy sound lasted only until 11 a.m. this morning when they decided to pack up and leave! What a state of affairs. Only another couple of days riveting is required on the rack when it could be lifted . . . *Feb 8* the situation over the riveters is quite unbelievable. Three more groups have actually collected their travel vouchers but have failed to arrive.

At last the riveters finished their job on the rack and in a snowstorm on 13 February there was the splendid sight of it being hoisted to the top of the southern tower. Another two weeks and more delays with riveting before the large girder frame was ready for the lift. On the

27th it was lifted on end, with the intention of hoisting on the 28th. Then a crane broke down and by the time it was repaired in early March there was a tremendous gale which stopped everything for days. We were worried, with only two ribs going half way to the centre ring on the scaffolding tower. '*March 1.* Husband says the telescope is in its most dangerous condition by far until the ribs are across.' By 5 March the gale had subsided enough for the great frame to be hoisted. By mid March the central two southern ribs had been fixed from this frame to the central ring of the bowl. '*March 19* . . . the second of the pair was raised today. It looks enormous now that the full parabola can be seen in position.'

Always the telescope building made apparent progress in great jumps as some big member was placed in position. Then there had to be consolidation, the filling in with the bracing members and the bolting and riveting. This was never more so than in the case of the bowl assembly. The great ribs were laid out on the ground, and it was after they were hoisted that the long task of the final connections between them took place. '*March 29*—the site packed up for Easter at 11.30 this morning having got further ahead than I ever thought possible a month ago. The central rib section is now virtually complete apart from riveting and crows are busily nesting amongst the trunnions.'

The vital issue now was to finish the bowl steelwork so that at least a start could be made with the membrane during summer weather. Unfortunately after the good progress in March there was a shortage of men and progress became very slow. When Renold came for his preliminary look on 12 April 'he was shattered by the little activity on the site and I got in a good 2 hours propaganda about the action necessary and the urgent need to get the bowl structure finished by June so that there was a chance of getting on the steel surface'.

Renold's site meetings were held in my office at Jodrell and at the first one on 20 April Husband produced his programme of work for completing the bowl framework by 7 July. 'Renold wrote to Commander Wells[1] inviting him over in a fortnight. This particular problem seems to be essentially one of getting enough man power on the site.'

May 5 (Saturday) . . . the first of the twelve secondary ribs is unfortunately still on the ground. Every day since Wednesday they have intended to lift it first to the vertical by No. 3 crane (one of the two smaller cranes which had been brought to the site to help in laying out the steel for the ribs on the

[1] Lt.-Cdr. G. W. Wells, D.L., M.I.Mech.E., R.N., then Chairman, Appleby-Frodingham Steel Co. Ltd., a major firm of the United Steel Companies, now a director of United Steels.

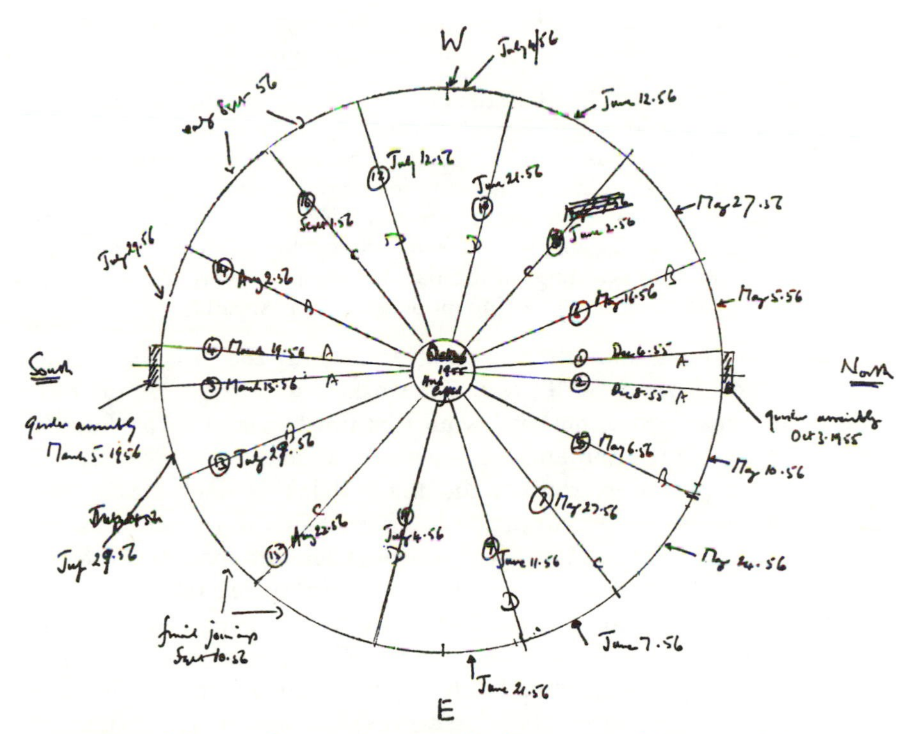

ground) and then to the top, but there is always some delay . . . this morning the first of the NW panels was lifted but all work ceased at 11.30 no doubt because of the cup final.

(The panels referred to are the large outer sections of the bowl structure going across the periphery of the ribs.)

Two days earlier Renold held his second site meeting. Wells couldn't come but Paterson[1] the general manager was present 'and damped us all by saying that Husband's programme for completion by July 7 was impossible. Their own date was the end of August and hence the end of September for the trueing up . . . the labour situation is unbelievably touchy and they are very much afraid of strikes.'

The first of the secondary ribs was in fact hoisted into position on the north-east side early on Sunday morning 6 May. Then the site became a giant jig-saw puzzle. The scaffolders were building their great towers around the railway track of the telescope so that each rib could

[1] Kenneth Paterson, the general manager of the United Steel Structural Co. Ltd.

be supported until it was bolted to its circumferential connecting panels and steelwork. Thus no rib could be lifted until its tower was ready. There was congestion on the site limiting the framework and ribs which could be built on the ground, and every movement of a crane had to be determined by a complex of timing and space. As the days of May lengthened there were increasing signs of progress.

May 15 . . . the rib was lifted off the ground today but a stiff breeze made it impossible to lift and it is now vertical on the ground ready to go up tomorrow. The riveters are also on the second E rib and it was hoped to get this up this week before Whitsun, but the scaffolders are too slow and won't have a tower ready until Thursday next. There has been a fuss about the rate of tower building and more effort is promised. The purlins are now in the central N sector and the beginnings of the paraboloid are apparent . . . with 4 cranes going, progress is at least being made with some speed.

At Renold's Site Committee meeting on 16 May the directors of the erection firm, Cozens & Butters, were asked to attend. A definite promise was made that the bowl would be finished and ready for the membrane in September and 10 August was accepted as a target. '. . . apart from labour troubles the main delay is likely to be the scaffolding towers, but urgent action has been taken to reduce the time per tower to 9 days . . . at last it really does look as though the bowl must be sheeted this autumn.' On that morning, before the Site Committee assembled, the first of the NW secondary ribs had been lifted. '*May 28*, the second NW panel and second NE rib went up yesterday. . . . now there is great activity. Twelve men are on the next scaffolding tower and riveters are at work on the bowl steelwork.' By early June all the filling-in steel and the purlins for the membrane were in position over a fifth of the bowl on the northern side. The paraboloid was taking shape and the work was ahead of the August target schedule.

June 11 . . . the last rib in the NE quadrant went up this morning . . . from the Bomish Lane entrance the telescope now looks terrific. At last one can begin to see what a 250 ft bowl looks like suspended 180 ft above ground . . . [and then] *June 14*. A black day. Came back from vivas in the University at 3 and found the men on strike . . . trouble over the bonuses.

After six days the men agreed to return to work while their dispute was subject to arbitration: 'it's all over a stupid business—whether the Whit Holidays should count as days for the bonus award.' Good followed bad. On 21 June the tenth rib and another panel were hoisted: '. . . hence the EW line is straddled and all the major components in the N hemisphere are now in position.'

Inspired by the sight of the erected framework for half of the bowl I began to press the question as to when we, as scientists, would be allowed to make a beginning with the installation of our own bits and pieces on the telescope. Hopes of having the telescope in a steerable condition before 1957 had vanished, but if the bowl membrane could be completed then there was much we could do even if it remained stationary in the zenith.

There was still a good deal of filling in of minor steelwork to be done in the northern part of the bowl and the first of the ribs in the southern half was not in position until 4 July. Erection procedures then became extremely complicated. The 62½-ft. aerial tower which had been built on the ground had to be raised while the cranes could still command the centre part of the bowl, and the construction of further support towers on the southern side could not be done because the cranes were in the way.

The large aerial tower was raised to its position first and by the time of the next site meeting on 25 July, the first of the western ribs in the southern sector was in position and the whole of that section completed. At that stage the erection was two weeks behind Husband's programme but two weeks ahead of the contractors. With four more ribs to go up the regular erection procession from north to south was altered in order to give room for the cranes to manoeuvre and the ribs adjoining the main southern tower were next on the schedule. The first of these on the eastern side was erected before the end of July and the one on the west side on 2 August. The cranes were then moved away from the structure so that the scaffolders could build the towers under the final ribs.

Aug 21 . . . telescope situation most aggravating . . . the East tower was finished on Friday; the panel and rib are ready but nothing has been lifted because they are dismantling the small crane . . . so the days go on and although we're probably within 10 to 14 days of the completion of the bowl steelwork nevertheless we're 4 weeks behind Husband's schedule of completion by August 8. . . . *Aug 22* another disturbing site meeting . . . raised hell about the work on dismantling the small cranes instead of erecting the bowl but Husband quite rightly pointed out that he could not instruct the United Steel Structural Co. on this matter without incurring risk of charges.

Actually the final rib in the eastern section was lifted on that day and the last rib of all in the western gap on 1 September.

The construction had, indeed, reached a dramatic state, with the final panels being built *in situ* instead of on the ground before lifting as a whole.

Sept 7. The E gap is joined out to the last cross piece and even the bottom one is across. On the W side the first cross braces near the middle are in position. Will it meet without trouble on the outside?!

Mon. Sept 10. The bowl of the telescope is ringed! The bottom level was joined this morning on the W side and now it's a matter of filling in the outer sections on E and W.—*Sept 12.* The last two big girders of the bowl went up this afternoon and both the final west and east top panel sections were completed within a few minutes of one another. Quite a moment! Now only minor pieces for the filling in and the big cranes have nearly finished their work. The derrick for dismantling them is here and so is the winch which is to lift the remaining pieces of the stabilizing girder and swinging laboratory.

The membrane

Pleasure and relief at the successful completion of the bowl framework were tempered by the knowledge that at least a vital month of summer weather had already been lost on the schedule for a start with the plating of the bowl. Although the steel was in position much levelling and riveting had to be carried out before the fixing of the membrane could begin. At the Site Committee on September 20

Husband tabled his new programme showing commencement of sheeting Oct 15, completion Jan 15. There was a long argument about when we could get on because the sequence indicates that the stabilizing girder and swinging lab won't be ready until Feb. March. Finally it was agreed that we could get our aerial on and have the head amplifiers on the winch platform.

Oct 2 . . . hold up now in riveting . . . the date of Oct 15 for beginning the membrane looks a bit sick. . . . *Oct 15* completion date [of riveting] receded by 2 weeks . . . however Husband said that they proposed to start plating on Oct 20 as soon as the centre rivets were done . . . *Oct 29* the purlins are nearly set out to 7 rings but the platers are not yet here . . . *Nov 6*—still no plating. Perfect calm sunny conditions and yet the welding gear is not yet in position.

As the hopes of getting a surfaced bowl by the end of the year receded I became rather annoyed. The same story which had so frequently bedevilled progress: never the real hazardous difficulties of erection, but always petty organizational delays. I was not the only one to be annoyed: '*Nov 8.* Husband is furious about the let down of Orthostyle. Their electrical supplies are still not at the top.'

At last on 12 November a start was made; two months behind the promised target made in May and nearly a month behind the target set in September when the bowl was complete. The delay was to prove disastrous to the telescope programme, although in the beginning the plating seemed to make good progress and optimism returned. '*Nov 13* . . . some plates at last welded. One man did 14 today which seemed fairly good work to me. Relief anyhow that they've started at last.'

The procedure was first that Goodall and his assistants checked and adjusted the level of the purlins and the welders could then 'tack' a series of plates on to these purlins and when sufficient were in position they would complete the welding. All these operations moved out from the centre and became increasingly difficult because of the steepness of the bowl. All together 7,100 steel sheets about 3 ft. × 3 ft. and $\frac{1}{12}$ in. thick had to be welded to form the complete reflector.

Nov 20 . . . Plates tacked out to purlin 9 but not welded. This final welding can only be done under dry conditions and so far this has only been possible on two days. *Nov 28* . . . still only one welder on site, the surface is now tacked out to circle 10 (there are 46 total) but very few are welded. Unfortunately this man will have to go because he belongs to the boiler makers union whereas the others belong to the constructional engineers union! . . . *Nov 30* . . . we now have the largest plated telescope. It was over 80 ft at lunch time and another ring is going on this afternoon. . . . the next trouble is that they will catch up the purlin setting which is still only out to purlin 23.

Alas! we were as far with the plating as we would be for many months. The dreadful forebodings of the effects of the summer days became a reality. At the site meeting on 4 December 'Husband read a series of letters from Orthostyle amounting to the fact that they proposed to abandon the job for a few months because of the weather. This is unspeakable particularly since the weather is mild and dry and balmy and plating is proceeding apace.'

On 23 December the welders packed up for Christmas and did not return.

Removal of the cranes and scaffolding

On 10 September, when the telescope bowl steelwork was finally joined around the outer circumference, the entire structure was encased in a vast mass of scaffolding. A tower supported the central hub and 16 circumferential towers 130 ft. high supported each of the ribs. These structures contained 90 miles of scaffold tubing weighing 750 tons, and the two cranes commanded the whole.

Now the work of the cranes and the scaffolding was nearly done. A number of miscellaneous items still had to be lifted: cladding for the tower tops, walkways, and ladders, but mainly the steelwork for the stabilizing girder and the swinging lab which was to hang beneath the bowl. Some of the stabilizing girder had been fitted earlier in the autumn and it was intended to lift the remainder and the other items by electric hoists instead of the cranes.

The dismantling of the cranes began early in October: '*Oct 29* [the cranes] are now down to gabbard level. This must be a devil of a job but another mobile crane is coming to assist soon.' The scaffolding also began to disappear. '*Nov 22* Yesterday about half the scaffolding towers were released [i.e. their support of the ribs released] and Goodall said the results were quite satisfactory.' A few days later and all were released. '*Nov 28* . . . the weight of the bowl is now on the bearings and this stage seems to have been passed satisfactorily.' By mid December the scaffolding was steadily disappearing but the central tower was still taking the weight of the central regions of the bowl and two days after Christmas I had to make a hard decision.

It was originally intended that the whole of the bowl structure should be supported on the temporary towers until the reflecting membrane had been completed. Since the steel sheets are welded to the purlins this would have added appreciably to the structural strength of the whole reflector framework as a stressed skin effect. Unfortunately the delay in the membrane and the refusal of Orthostyle to proceed during the winter led to a critical situation. As long as the scaffolding was present, particularly that supporting the central hub, it was impossible to proceed with the stabilizing girder, swinging lab, and various other parts of the telescope. 'Husband phoned . . . because of the plating hiatus they want to remove the central scaffolding tower. . . . Calculations show that this will allow the bowl to distort so that the focus relapses about 2½ ins inside the aperture, with a corresponding different parabola.'

If the background to my anxiety about the completion date of the telescope had the slightest foundation I had little alternative to agreeing instantly to this proposal. Equivocation would have removed all possibility of completion until the winter of 1957.

25

Finality on the driving system

THE long saga on the driving system has been described in Chapters 16 and 20. At last at the beginning of April 1955 we seemed to be on the point of placing the orders for the driving system and gearing with Messrs. Siemens in Germany. But with the sudden revelation of our financial plight, all action ceased on that matter. Apart from one or two tentative high-level approaches to The Company to probe the possibility of a gift of the driving system nothing was done for a whole year. Then, with the D.S.I.R. Advisory Council approval of the report of the inquiry and of the decision in principle to pay half the deficit, urgent attention was once more given to this problem, even though the Treasury's approval had not been received. At Renold's first Site Committee meeting on 20 April 1956 'Husband was instructed to get new prices and delivery dates out of Siemens urgently'.

At the second meeting on 3 May Husband had the revised figures:

> they were only 10 per cent over the original ones and the delivery promise remained 8 to 9 months. This will certainly be accepted but is to be held up for a week because Brush are anxious to put in a tender. Their only hope is that we shall have to pay import duty on the Siemens equipment. . . . *May 15* Cannon phoned to say that the Brush estimate on the driving system was unlikely to be acceptable.

It was not—at least to us—and at the site meeting on 16 May it was agreed to accept the Siemens tender and Rainford was asked to obtain the necessary permission from D.S.I.R. to place the contract.

We thought that this would be a formality. In fact the Siemens' contract was never placed. D.S.I.R. said that they must first see the other estimates and Husband had to be asked to draw up another report. '*June 6.* Rainford said that the documents about the driving system were with D.S.I.R. The delay in ordering is infuriating.' On 12 June I saw Greenall in London who told me that D.S.I.R. would give approval to the Siemens contract with some 'reservations about the method of payment'. At a meeting of the University Officers on

15 June Rainford obtained permission to place the contract in spite of the grim financial state and by 20 June we understood that Husband had sent a formal acceptance of the contract. Then at the site meeting on 23 June:

a most extraordinary situation over the driving system. We understood last Thursday that Husband had acted on instructions to place the contract for the driving system with Siemens. He had all letters written and ready for dispatch when Brush phoned to say that they could now reduce their figure to a competitive level with roughly the same delivery date as Siemens . . . The overall estimate was still in excess of Siemens but would be less if the import duty had to be paid. Rainford phoned D.S.I.R. and after a short time he got permission to place the orders with Brush if he were satisfied with their competence. This will now be done!

I never found out what pressures and intrigues took place between 3 May and 23 June which ended in this last-minute switch of the contract from Siemens to Brush. I was just too thankful that at last this devilish issue had been resolved and my own worry concerned the apparent lack of experience of the English firm in this form of drive. In the event we had our troubles but they were always of a minor character and the extraordinary reliability and efficiency of the Brush system was to give us prolonged satisfaction.

The arrangement with Brush was that Messrs. Wiseman of Birmingham should manufacture the major gear trains required in the azimuth and elevation drives.

Sept 25. There is trouble about the gears. Although Husband says that the Brush dates are OK, Wisemans give 12 months. It was agreed to get them here for a talk and also to explore the possibility of fixing a definite opening date of July 1st 1957 . . . the Brush contracts were sealed by the University Council on Friday even though they haven't got the money to pay. Hence, at last, all major contracts are placed!

On 8 November the Chief Executive of Brush and the directors of Wisemans attended the site meeting: 'the hold up is the gearing. Brush's dates are for next March but Wiseman's best is May . . . Ways and means of improving Wiseman's dates by a month were considered and the necessary liaison set up.'

Although I had pressed repeatedly for some data on the Brush system it did not come. Eventually J. G. Davies and a colleague from the Electrical Engineering Department of the University who was a specialist in control systems visited Loughborough and in late November I was glad to be reassured that the proposed servo system would meet our requirements. By the end of the year the news from Brush and

Wiseman was good, the urgency which had been conveyed to the directors on 8 November bore fruit and some means had been found to bring forward the dates of the crucial gearing. We were promised the delivery of all gears and machinery by April with the expectation that assembly would be complete by the end of May.

26

The Treasury and the Public Accounts Committee

THE year 1957 and the last months of the telescope construction should have been full of the joy of achievement. They were, in fact, bedevilled by a sequence of events of which I could be only an anguished spectator.

After the meeting of the D.S.I.R. Council in March 1956 which considered the report of the Committee of Inquiry we anticipated that our punishment would be the raising of half the outstanding sum amounting to £130,000. The Treasury still had to give approval to the D.S.I.R. decision and, as mentioned at the end of Chapter 23, our expectation that this was to be merely a formality turned out to be a grievous misreading of affairs. On 9 April 1956 in my notes on the meeting with Lord Simon and the other officers to decide Sir Charles Renold's terms of reference (Chapter 24) I noted that 'there is one small snag—the formal Treasury approval has not yet been obtained for the D.S.I.R. £130,000 contribution but Rainford isn't worried about this and expects it any day'. It was rare for the Bursar not to be worried about a financial outlook, but on this one occasion he was, unfortunately, wrong in his expectation of the early Treasury approval. '*April 20* . . . the financial position is still uncertain. The latest news from D.S.I.R. is that the Lord President is being awkward. . . . *May 25* . . . still no news of Treasury approval for the additional D.S.I.R. grant. Rainford pressed this last week in D.S.I.R. It is believed that the Treasury is afraid of the Public Accounts Committee.'

In June affairs began to take on an ugly look. The Treasury were said to be asking why the over-expenditure could not be recovered by legal action; '. . . the belief is that they are very much afraid of the Public Accounts Committee.' The University Officers were repelled by the suggestion of legal action and sought Counsel's advice. '*June 8* . . . the ugly fact remains that we still have to find £260,000 and that the Treasury are still haggling about their half of it.'

We had other troubles, too. The Committee of Inquiry had accepted a figure of £650,000 as the final estimate, leading to the deficit of

£260,000 to be raised. At the site meeting on 3 May Paterson from the United Steel Structural Co.

> also raised the question of the cost of the structure and there was some dismay but it seems that most of his points must have been considered already by Husband in the figure given to the Committee. . . . *May 25*. Cannon said that Husband had sent on a letter from the United Steel Structural Co. stating their claim for an additional £117,000 over the agreed steel tonnage price of £114 per ton . . . Husband says this cannot be allowed and has expressed confidence that the Committee's investigation will have yielded a nearly correct figure.

The position was one of enormous complication, and when this claim was discussed at length by Renold's Committee on 8 June Rainford and Cannon thought that only about half of this was included in the Committee of Inquiry's figure. We began to wonder even if £260,000 represented the limit of our liability.

On 13 June in London I saw Greenall and Smith and they again said that their view was simply that the Treasury wanted to cover themselves fully against the Public Accounts Committee and questions in the House:

> . . . D.S.I.R. obviously takes a poor view of this Treasury intervention in handling money which has already been granted to them . . . *July 25* . . . no progress whatever with the Treasury and Rainford and the Officers are obviously getting very anxious . . . *August 1* . . . the horrible situation now seems to be arising that the Treasury may not give their permission unless the University takes legal action . . . this is of course fantastic; an opinion which according to Rainford was shared by the Officers.

By mid August we were given to understand that the Treasury decision had been made and that D.S.I.R. would be sending a 'try out' letter to Rainford. 'It looks as though a series of nasty conditions are to be attached.' They were: '[*August 15*] . . . Greenall phoned this afternoon with news about the grant. At last the Treasury had given an answer. They are willing for the D.S.I.R. to make half the additional grant i.e. £130,000 but there are 4 pages of conditions which stick in D.S.I.R.'s throat and which Greenall is afraid will stick with the University as well.' The worst condition was that the University had to give a written guarantee, with Counsel's opinion, that the sum was not greater than it would otherwise be if legal action for recovery of the money had been taken.

The Treasury had insulted the University, and all concerned with this great project, which should have been—and indeed was soon to become—the country's pride. Of course, I have no knowledge of the internal considerations which motivated the individuals in the Treasury who remained unnamed and unknown to me. Our external assessment that it was fear of the Public Accounts Committee would appear to be correct—as will be seen from what followed. I felt at the time and still feel today that these conditions were made by men whose first concern was the preservation of their own skins, who had no concept of the likely consequences of their action, and whose horizon was limited by the walls of their offices. It must be remembered that there was no question of the Treasury providing additional money since on the advice of the Committee of Inquiry and of their Advisory Council, D.S.I.R. intended to find the money from sums already voted to them by the Treasury. The consequences of the Treasury's conditions were disastrous in the extreme, and only by the narrowest of margins did the project continue to move forward to success. The episodes which followed nearly ended my own career and that of the others principally associated with the telescope.

It was impossible for the University to accept the Treasury conditions and the instant reaction was to explore other avenues for the money. It was suggested that I made private approaches to some of my American colleagues, which I did immediately. I also conveyed to certain people in London who had an interest in the early use of the telescope that the Treasury conditions were incompatible with its rapid completion. Then, as so often happened after a violent disturbance, there was a long period of quiescence during which the building of the telescope continued. '*Oct 29* . . . on the financial front things seem static. Rainford is seeing Counsel in London tomorrow. Blackett was here on Saturday and seemed pretty pleased with the look of the telescope. Since nothing is now held up for money we agreed to let things go on for the time being.' The year 1956 continued and ended with the University in constant touch with its Counsel and D.S.I.R. in an attempt to find some formula by which the Treasury conditions could be made acceptable.

The Public Accounts Committee

In 1954 Sir Ben Lockspeiser, who was the Secretary of the D.S.I.R., had been questioned about the expenditure on the telescope by the Public Accounts Committee. It was not then a major issue and D.S.I.R. was at that time facing graver issues before the P.A.C. on the question of a serious underestimate on the costs of a fisheries research trawler. I have mentioned in Chapter 18 that Sir Ben offered to bring the members of the P.A.C. to Jodrell Bank and this resulted in his visit with Mr. Benson (later Sir George Benson), who was then Chairman of the P.A.C., on 20 September 1954. Since Lockspeiser had at that time 'explained to them [the P.A.C.] that no more money would be available for the radio telescope' it was to be expected that the Committee would take notice of any future suggestion to supplement the D.S.I.R.'s contribution.

Even so, the sum of money involved was trifling compared with the discrepancies which the P.A.C. normally investigated (the investigation of the £260,000 telescope debit, appeared in the White Paper next to an investigation of a wastage of several million pounds on Army boots). It is therefore unlikely that the telescope finances would have assumed the prominence and importance which they did had it not been that they were associated with other major issues of policy.

It was our misfortune that the telescope was perhaps the largest and most expensive single item of research equipment that had, at that time, been undertaken by a University. It was also an accident of timing that the period was one in which voices were being raised against the traditional freedom of the Universities. The sums being provided by the Exchequer for University purposes were soaring rapidly. The grants to the Universities came directly from the Exchequer to the Universities through the University Grants Committee and were not routed through a ministerial department (as they have been since 1964). The University books were not therefore open to inspection[1] and in spite of the considerable amount of public money voted for their use, the Universities were effectively free agents in the use they made of their grants.

Strident voices were already being raised in the P.A.C. about these matters and their inability to criticize the University expenditures. Thus the relatively minor financial troubles of the telescope came as a godsend to those voracious individuals who sought for means of gaining public control of University affairs. Every aspect of the matter was seized on as indicating that Universities were incapable of controlling the expenditure of large amounts of money. This was the background to the anxiety of the Treasury and the University officials. On 4 January 1957 Rainford said at a site meeting that the financial situation remained 'terrible' and no one would move for fear of the P.A.C.

> Afterwards I complained to Renold about the difficulty of finding out the exact position and he agreed. One is constantly confronted by high sounding phrases about the freedom of the Universities being at stake. . . . *Feb 6* . . . the over-expenditure remains a mess but Rainford said he intended to wait now for a time—the trouble is entirely P.A.C. and issues of the Treasury control of University finance are at stake.

[1] But in the summer of 1967 the Labour Government passed legislation ending this favourable state of affairs.

I am in no position to assess what effect, if any, the finances of the telescope had on the transfer of responsibility for University finances which was made seven years later, but our own immediate troubles were soon to come to a head. Fortunately, by the time the crisis came the completion of the telescope was inevitable, and indeed, not for the first time, it seems that the tardy movements of the financial machinery saved the entire project.

Sir Ben Lockspeiser had retired during this initial phase of the telescope's affairs and his successor, Sir Harry Melville,[1] had the unenviable task of appearing before the P.A.C. to which he was summoned on 21 March. On 19 February, as part of his preparation for this task, Melville came to Jodrell Bank with several members of the Committee of Inquiry in order to familiarize himself with the telescope and its background. The appointment of Melville continued the succession of distinguished scientists who held the Secretaryship of the D.S.I.R. and it was a relief to spend the afternoon discussing the astronomical aspects of the telescope instead of its finances.

Unfortunately on 5 March Sir Frank Tribe's report was published and this document contained the straight factual story of the increasing cost of the telescope. The influence of the Press on our affairs remains to be told (see Chapter 29); at this juncture it is sufficient to remark that apart from an unconfirmed rumour which appeared in a Sunday newspaper in April 1956 there had been no public indication that we had accumulated a debt of more than a quarter of a million pounds. '*March 6* . . . The newspapers are having a field day.' The timing could not have been worse. The Vice-Chancellor and the Treasurer had only just completed some satisfactory negotiations with Husband who had been most generous in his attitude to his professional fees. He was understandably furious at this unexpected release of an unbalanced part of the story—unbalanced because it was essentially factual without explanation of the attendant circumstances.

Throughout that spring, as the telescope moved to completion, the background was tense with apprehension. Legal aspects predominated and I became afraid to write letters on anything but the most essential practical matters. I continued my diary of the telescope but it was no longer typed. '*April 18* . . . yesterday Hingston[2] came with further worries. Although he has no precise information he believes that the

[1] Sir Harry Melville, K.C.B., F.R.S., had been Professor of Chemistry in the University of Birmingham since 1948. He remained with D.S.I.R. and then (in 1965) as Chairman of its successor, the Science Research Council, until 1967 when he resigned to become Principal of Queen Mary College, University of London.

[2] Col. Walter G. Hingston, Public Relations Officer in D.S.I.R. Headquarters.

P.A.C. report will be strongly critical.' Also, at last the Treasury's conditions for agreeing to payment of half of the grant were officially transmitted to the University by D.S.I.R. '*July 1.* Lord Simon phoned on Sunday morning. He is seeing the Head of the Treasury tomorrow to try to get these ridiculous conditions removed. It would really be fantastic if the University was forced into any legal action.'

The P.A.C. report which caused us such gloomy forebodings throughout those summer days finally appeared on the afternoon of 13 August. We had reached our nadir: '*August 16.* The dreaded P.A.C. report came out last Tuesday afternoon and the worst 24 hours of the whole telescope followed. The report criticized Husband for making substantial changes without consulting me. The newspapers interpreted this as lack of general cooperation.'

In our worst moment I do not think that any of us expected this personal attack. The interpretation was, of course, utter nonsense: for the previous seven years scarcely a day had passed without a personal, mail, or phone contact between me and Husband. Husband was understandably very angry indeed, '. . . phoned in the evening and asked me to write a letter to *The Times* denying these allegations otherwise he would sue me personally'. I replied that the P.A.C. report was a privileged document and that I could not possibly move without the permission of the Vice-Chancellor. He was on holiday and Rainford said that I must on no account write such a letter.

'The irony of this situation is quite unbelievable. To be slashed by a government committee when this marvellous instrument is beginning to work is a most strange business.'

27

The telescope moves

THE publication of this P.A.C. report did, indeed, coincide with the moment when we were beginning to test the telescope. 1957 opened with a farcical situation. Orthostyle had left the site before Christmas with only 80 ft. diameter of the bowl plated and refused to return until the weather improved. The year began and continued like springtime, the weather conditions were far better than those which the Orthostyle welders had worked through in the autumn and yet they refused to return to the job. '*Jan 3* . . . I have sent Husband our rainfall records for July–August, November–December which show far worse conditions in the summer than those which existed when Orthostyle were on the site in November and December.' The steelworkers were on strike and the sole activity on the site was the dismantling of the scaffolding.

I did not, of course, have any official contact with the workmen, although since I studied nearly every bolt and rivet that went in we were well acquainted. Even so I was surprised on 3 January to be visited by a deputation of the Shop Steward, his deputy, and a man representing the erectors. 'They wanted to put their point of view about the strike which was essentially, that on Dec 31 they had been summarily informed that the bonus arrangements would not apply as long as they were working on the purlins. They complained that alterations had become necessary because of errors in drawing.' Their argument was that they were being made to suffer because of a dispute between the contractors as to who should pay for the extra work involved.

I could not enter into these labour problems and immediately passed the issue on to the Site Committee which met later in the afternoon. 'The Site Committee was, as usual, helpless in the face of the strike and the refusal of Orthostyle to continue with the plating until the weather improved.'

The strikers came back seven days later and agreed to work while negotiations were in progress. It was one of the driest and mildest Januarys on record and before the end of the month the adjustment and setting of the purlins had been completed. '*Jan 21* . . . wrote to Renold pointing out that the steel people had lost less than 2 days in the first 18 of January on account of bad weather and saying that the decision to do nothing about Orthostyle was convenient administratively but was inconsistent with the spirit in which the University had brought the telescope to its present state.'

Husband was anxious to measure the azimuth track friction and on Sunday afternoon, 3 February, we had the excitement of seeing the telescope move:

Only an inch, but everything was free. The central scaffolding tower was released on Friday and is now being taken down as quickly as possible, and the remaining supports on the ring towers were burnt away on Sunday morning. The bowl rocked rather gently in the wind while the 50 ton hydraulic jacks were applied to the bogies. . . . finally Perry and Goodall had to climb up the structure to wedge the bowl . . . *Febr 6* . . . for the first time I really felt that the telescope was moving inevitably to a finish in the summer.

The Orthostyle management was summoned to a site meeting and 'gave reasons for not starting again until April. However he promised to finish by mid-June at the latest.'

Meanwhile the remaining pieces of the steelwork, the stabilizing girder, the cladding on the elevated motor rooms, and the swinging laboratory, steadily went into place. Early in March the second crane lifted its last load and dismantling began. Then on 30 March everybody went on strike again. '*March 30* . . . what is particularly infuriating is that the Orthostyle people turned up on Thursday ready to start on Monday as promised and were ordered off the site by the Union officials.' Another vital week was lost.

These troubles were forgotten by mid April when there was 'tremendous activity on site. A lift actually works and yesterday afternoon I ascended in it with Kington'[1] (16 April). The Wiseman gearing and the Brush motors began to be delivered and the riveters had started on the stabilizing girder. The membrane was still the holding item; 'the platers after over a week are still correcting the distortions in their original welding'. By this time we were in the midst of cabling problems. The 'domestic' cables for the power and lighting on the structure and the cables required for the research apparatus were complex and we had grossly underestimated the cost of this work.

April 10. . . . violent arguments with Cannon and Hanson.[2] I consider that the only thing is to treat the excess £8000 on the cabling in the same way as our £8000 radio frequency cables, i.e. as a research item . . . *April 16* . . .

[1] C. N. Kington, M.B.E., who had recently joined Husband. Subsequently he assumed a major responsibility for the mechanical aspects of this telescope and its successors at Jodrell Bank.

[2] James Hanson, the Finance Officer in the Bursar's department.

leaves the £8,500 excess on Section 1 [the domestic cables] and the £8,400 radio frequency Section 2 cables. The suggestion that the University might advance this as a mortgage on my research grant was the subject of an estimates and stipends Committee this morning to which I was summoned. The Vice-Chancellor and the Treasurer would not buy it.

However, everybody realized that it was now useless to hold up the essential power wiring and early in May the orders for the Ward Leonard cabling were given. I was left to beg as much as possible of the research cabling and the fact that we were able to work at all later in the summer is a tribute to the generosity of one or two of the major cable manufacturing firms.

Orthostyle failed to keep their promises. Early in May I was again complaining bitterly: the whole of April had been occupied in dealing with the autumn plating and not until 5 May was the area of the plated bowl increased from its pre-Christmas size. Eventually the men were put on piece work 'which, of course, is the essential issue . . . Husband still maintains that this plating could be completed by June 30 which remains the target date' (7 May). Indeed the effect of introducing piece working was a revelation (to me at least): three weeks later (May 27) '. . . spectacular results and there are now only 17 rows left to be done —one counts the rows undone instead of those already done'.

One feature of these days was the remarkable way in which Brush and Wiseman kept their promises. The driving system which had bedevilled the telescope progress for years eventually appeared as if by magic. '(*May 27*) . . . Brush and Wiseman have done a marvellous job and now the motor and gears are going on the driving bogies. The Ward Leonard room is nearly complete, both motor generator sets and even some of the electronic equipment is installed.'

By the end of May all the cranes had vanished from the site, the painters were hard at work and the riveters, too, although they had little remaining to be completed. The bowl was still supported on two scaffolding towers; the swinging laboratory had been hoisted into place underneath the bowl and all the walkways were complete. 'The atmosphere is tense and exhausting. The publicity issues are beyond belief.'

Azimuth motion under power

Indeed by this time I was myself nearly prostrate with exhaustion. The years of difficulty and tenseness almost without intermission and relief had taken their toll. Not only the telescope, but the researches at Jodrell, and our commitments to the International Geophysical Year were part of the daily run. On 1 June I went with the family to North

Wales. I returned on the 6th for a site meeting 'the day after some fool had overloaded all our generators and stopped the station 3 hours after the first IGY alert'. I returned again to North Wales and on 12 June climbed Snowdon with the three elder children. We returned to find the remainder of the family in an excited state: Husband had phoned from Jodrell to say that the telescope had moved in azimuth under power.

Wednesday 12.6.57

Dear Professor Lovell,

We have just seen the telescope move 20 yards—it was most exciting and the suspense was terrific! In fact at this moment it is being driven back to its original position. Everything seemed to go very smoothly.

Enclosed are the memorandum and registered envelope.

Hope the weather holds.

Yours sincerely,
Maureen Patrick[1]

Another week and I was back to find the telescope 'in handsome condition. No further movements but everybody thrilled with the way it moved last week. Less than 50 kw on the diesel and Davies said the progression from no movement to motion was superb . . . we went inside the bowl which is quite fantastic. They still have about 12 circles to do, not much but nearly one quarter of the area!' (18 June).

The bowl moves under power

We were close to the greatest moment of all:

Thursday June 20. 57. The bowl has been tilted! A marvellous summer morning and after much clearance of the site and general preparations the bowl began to move at about 10.30. The transition from no movement to movement was imperceptible, the motion was superb—only a quiet hum from the motor room like a great ship. After about 10 degrees they stopped to regrease the gear racks and then at about 12 o'clock it was put right over to 30 degrees. A majestic sight in the hot sun.

At least for a moment all the past agonies, the present and future troubles were forgotten—

21 June 1957

Dear Bernard,

Very many congratulations! This is great news. I will look forward to visiting you and seeing the telescope some time next month . . .

Yours,
Patrick (Blackett)

[1] My secretary at that time, who subsequently married one of my research students, Dr. J. V. Evans.

A memorable bank holiday—the first signals

Although the bowl had been tilted there was much still to be accomplished. The membrane was still unfinished. '*July 1* . . . The plating of the bowl is out to the last circle on the S-side—another 2 or 3 weeks Husband thinks . . . Dunford and Elliott are still in the control room but dials are now rotating. . . . *July 17* . . . things have reached the dragging on stage . . . yesterday they [the platers] started the last row and by lunch time were about ¼ around, then it rained again. The electricians are not even on the bowl yet and the official Brush tests still have to begin.'

At last we infiltrated to the extent of getting an aerial on the mast with a cable connecting it to a receiver.

Aug 7 . . . we shan't forget August Bank Holiday 1957. On the Friday night Aug 2, we got our first recording. 160 mc/s with the bowl in the zenith. A remarkable run completely free from any sign of interference although the galaxy was half scale. Hazard reckoned that the sensitivity was at least 6 times up on the transit telescope and Hanbury[1] said it was the finest record he'd ever seen. The next two nights this was repeated at even higher gain and also with the 90 mc/s on cross polarisation, but this is blacked out by the B.B.C. except for a few hours in the middle of the night.

These recordings were made although there were no cables available for our use. In fact our instruments were connected by cables coming straight to ground from the swinging laboratory. Gradually we tested more and more of the telescope. We were, of course, impatient to measure the polar diagram of the reflector on various frequencies. We began on 9 August. The full controls were not tested and we could only do these measurements by having the resident engineer in the Ward Leonard room on the telescope. There were no degree calibrations for the azimuth or elevation motions, so we measured around the track and the circumference of the stabilizing girder and made our own temporary fiducial marks. The movement of the bowl from the zenith was still restricted and we could only work after midnight when the radio source in Cassiopeia was sufficiently high in the sky. With field telephones and much shouting we measured the polar diagram on a number of frequencies and everything was fine.

Aug 23 . . . a note before leaving for URSI in Boulder. On Monday night we carried out a further series of tests . . . results are superb—again exactly as calculated . . . the processes of clearing up are slow. No further elevation motions yet or linkage with the control room. By the time we return I hope that many of these (and the painters and electricians) will be out of the way.

[1] R. Hanbury Brown, see footnote p.91. His young collaborator, Dr. Cyril Hazard, one of our students, achieved distinction early in 1960 for his discoveries relating to quasars in Australia and America.

They were not. '*Sept 17* . . . not very impressed at the rate of progress on the telescope. In fact I'm dismayed after a few days' investigation at the way in which the job is slowing down to zero. During our absence the bowl was tilted up once and this bent some girders, since when it's been securely fixed in the zenith.' Worst of all there were no signs of the tests of the Brush control equipment which had been scheduled for completion by 21 August. We had expected to find a fully steerable telescope at our disposal. We were far from it and arranged a programme of measurements which we could do with the telescope in fixed positions when no workmen were on it. '*Oct 2* . . . telescope atmosphere exceedingly disturbed since my return. At the site meeting [Sept 24] it was said that 14 days intense work was required to link up the driving system with the control room.' After seven days Brush appeared and succeeded in driving it backwards; 'due to come again on Tuesday but P refused to come back with "University people breathing down his neck".' Naturally this led to a vicious exchange; meanwhile the lifts were breaking down continually, the central clamping device still had not arrived, electricians and painters were all over the telescope and it still seemed that ages separated us from any reasonable use of the telescope for scientific purposes.

Two days later the Russians launched Sputnik I.

28

Public relations—authorities

SPUTNIK I and its successors had a profound effect on our affairs. On 3 October 1957 we had the attitudes and characteristics of the traditional scientific establishment, resentful of publicity and waging a war with the Press whose unwelcome attention we had attracted because of the size of our scientific instrument and its financial troubles. In the course of a few days we were transported on an irresistible flood to the front pages of the world's newspapers and had to adapt ourselves to a new life in which our work came under public scrutiny for long periods and in which the Press became our friends and fought our cause. It is convenient to pause at this stage in order to give an account of the evolution of our relations with the Press and public authorities during the construction of the telescope. First, our relations with the Ministry of Housing and Local Government, the Ministry of Fuel and Power, and the Postmaster-General.

The planners

When the Area Planning Officer first visited Jodrell Bank on 5 June 1952 he said that he anticipated no difficulties whatsoever in the granting of planning permission for the telescope. The difficulties which did exist were of our own making. As related in Chapter 9, at that time we had not reached a final decision whether to build the telescope in Field 80 or 132. Although adjacent, the two fields were in different rural districts.

In the light of the subsequent influence of the telescope on the development of this region of Cheshire it is a most remarkable fact that once we had decided on the field, planning approval was given almost instantly (there was a few days' delay because of a sewage disposal problem), and in early August of 1952 we received the official communication containing only three conditions: (1) we must preserve the trees and replant, (2) we must discuss the colour of the telescope with the Council for the Preservation of Rural England, (3) we must inform the Air Ministry (a) when the construction began and (b) when it was finished. I was puzzled by this last condition. Many years later when I was trying to establish the extent of our statutory obligations to have aircraft warning lights on the telescope I found the answer. Because of the height of the structure the planning documents had been referred to the Air Ministry. Someone there had looked at the maps and decided that the telescope was only a few miles from an aerodrome at which a night bomber squadron was stationed. Fortunately, the telescope was only likely to disturb the sheep which had long since taken over the decrepit hangars and runways.

A few months after planning permission was given I became aware through the local newspapers that the area planners were showing a great interest in the regions around the telescope as a site for a new town, or alternatively the enlargement of one or more of the existing country towns to deal with Manchester's overspill population. I had no conception then of the difficulties and massive confrontations with the public authorities which were to be enacted. Although the radio telescope could penetrate the thickest cloud, its work would be subject to serious hindrance in the presence of electrical sparks and other forms of interference generated by extensive communities in the vicinity. At that time Jodrell Bank was quite isolated. There were no electrical supplies even in the surrounding villages and we were free of any significant centres of population for at least six miles.

Seven miles to the south-west were the salt towns of Sandbach and Middlewich and the danger of subsidence was so great that no one entertained the idea of extensive building there. At the same distance to the south-east was Congleton. It became and remains a great anxiety to ourselves, the city fathers, and the planners. Envious eyes were cast on this region by the planners: the Congleton Borough Council wanted development, but to us it was vital that Congleton should not grow a great conurbation towards the telescope.

Nov 4 1952 . . . the recent news about the development of Congleton to house Manchester's overspill under the new towns act has again caused me some concern . . . phoned Cross [the Clerk of the Rural District] who could not give me the answers I wanted [about the extent of the proposed penetration]. He suggested I make representation to the Ministry.

I did, and so began the long haul.

Eleven months later it was Cross who was seeking my assistance on this problem. The Rural District Council had become alarmed at the possibility of Congleton's expansion being so great that a considerable section of the rural area would be absorbed, and the R.D.C. began to realize that the telescope might be an ally for them. As usual in these circumstances, the complications were endless. Manchester wished to extend its territory by absorbing Mobberley and Lymm for rehousing its people. The Cheshire County Council were violently opposed to this absorption of its territory and in cooperation with the Congleton Borough Council they were offering Congleton as an alternative site. The official University interests were double-sided: on the one hand they desired to help me to protect the telescope; on the other hand they could not realistically oppose any Manchester plans for overspill because of their urgent need for extending the University premises which could not be done until the people were rehoused. In October 1953 the Minister was called upon to adjudicate at a public inquiry as to whether the Cheshire objections to the Mobberley–Lymm proposals should be upheld. '*Sept 24* . . . seems to me to be the fourth successive autumn in which trouble has been stirred up over the rehousing of Manchester's population . . . long discussion with Cross as to the possible avenues through which we might help.'

I had already been in touch on a previous occasion with the Ministry of Housing officials but because of the delicacy of the situation I felt that it was essential to work through the D.S.I.R. machinery. Unfortunately my contacts there were on holiday and I sought Hingston's advice.

Sept 29. Hingston phoned . . . he had been in contact with B at the Ministry who had emphasized that the inquiry concerned only the proposals for housing Manchester's overspill in Mobberley and Lymm and that Congleton was not in question. I said that Congleton was certainly an alternative which Chester would be pressing against Mobberley and Lymm. However B said that the Chairman of the Inquiry Committee had not yet been appointed and that in any case it would be inappropriate for him to receive representations before the inquiry. He recommended that I establish contact with the Ministry's Regional Officer in Manchester . . . additionally dismayed last night to read in the annual report of the C.P.R.E. that they were opposing the Mobberley-Lymm proposal and were strongly in favour of Congleton.

October was thick with activity and meetings on this problem. The M.P.s of the district, the Chairman of the County Council, and the Regional representatives all separately came to Jodrell for discussion. We found that the Minister's regional representative was tremendously interested in astronomy and it was refreshing to find someone in that position who was prepared to regard the affairs of the telescope as of 'paramount importance'. Not everyone took this view of the telescope.

Nov. 10 1953 . . . telephone call from the Evening Chronicle giving the information that on Tuesday next Col. Erroll,[1] the M.P. for Altrincham 'would ask the Parliamentary Secretary to the Ministry of Works, represent-

[1] Col. F J. Erroll was the Conservative M.P. for Altrincham and Sale 1945–64. He was President of the Board of Trade 1961–63, Minister of Power 1963–64, and was raised to the peerage as Lord Erroll of Hale in 1964.

ing the Lord President of the Council whether he would instruct the D.S.I.R. to withhold any further grant for the construction of the radio telescope which is being built south of Manchester until he is satisfied that the public money so expended will not be wasted by satellite town development, seriously affecting the value of the telescope'.

I had never met Col. Erroll and failed to understand how he could assess the importance of the telescope in relation to new town development. We were in our financial crisis and the knowledge that he had visited D.S.I.R. a few days ago to inquire about the telescope finances increased my worries. Greenall phoned and we drafted the reply: 'No sir, it is in the national interest that the telescope should be built as quickly as possible and the only development likely to effect this which has been brought to my notice is the large scale expansion of Congleton to the north west.'

The next day, 11 November, I phoned Col. Erroll's agent to express surprise and annoyance at the tabling of this question without consultation. Later, after he had spoken to Col. Erroll, he phoned to say that Col. Erroll 'only wished to make sure that public money would not be wasted'. He did not seem to appreciate that the question should have been put the other way round, so I sent a telegram to my own M.P. requesting him to ask a supplementary: 'Will the Minister ensure that in the national interest the radio telescope be completed as a matter of urgency and that no satellite town development be allowed to prejudice its performance?'

In the interval of a Hallé concert in the evening I met H [the Minister's regional representative] and he was full of the trouble that Col. Erroll's question is likely to cause. Fortunately they had inquiries from their own Ministry and had replied in the same vein as the draft which I had agreed with Greenall.

Nov 14, 1953. In London . . . saw Greenall who gave me all the details of the reply and associated information now in the hands of the Parliamentary Secretary to enable him to deal with Erroll's question on Tuesday. The P.S. has been primed that this is a political manoeuvre because of local objections to the new towns in Mobberley and Lymm. The reply is a refusal to tamper with the grant and, what is more important, the P.S. will say that he understands from his Rt. Hon. friend the Minister of Housing and Local Government that should the occasion ever arise when the expansion of Congleton has to be considered then he is confident that arrangements can be made so as not to make any interference with the radio telescope . . . The Min. of Housing's department has therefore co-operated excellently in this matter and the Lord President's office has accepted the reply which is to be sent.

Nov 17. Erroll's question due this afternoon . . . 5.30 p.m. phoned Greenall whose only information was that Erroll had balloted a number over 100 and hence it was extremely unlikely that the question would actually be asked in the House.

The question and answer duly appeared in *Hansard.* It did not escape the eagle eye of Lady Simon: 'Is this really all right, or are you still uneasy?'

On 4 December 1953 the permanent officers of the Cheshire County Council, including the Clerk and the Planning Officer, made their first visit to Jodrell Bank to find out what our problems were and why we were concerned about housing developments. In spite of the troubles which we caused the county then and have continued to do in the following years, our relations with them have always been most friendly. It is fortunate that this has been the case; the good relations were to suffer many tests and a breakdown would have been, and indeed still could be, a serious danger to the functioning of the telescope. In those days our own position was weak. The telescope was scarcely above ground level and we were already in debt. At no time did the county or ministerial officials seek to take advantage of our difficulties to press their own case beyond their normal professional judgement.

At the time of that December meeting the county planners were concerned about our objections to Congleton but they did not treat it as a major issue, since it was their opinion that the Minister's decision after the recent inquiry would be in favour of the Mobberley-Lymm proposal. They were wrong in this belief, although it was to be many months before they knew this.

Oct 22, 1954 . . . At the Vice-Chancellor's reception on Wednesday night Lady Simon approached me in a most gloomy fashion about Manchester's overspill problem. A week ago a letter was received from the Minister of Housing [Mr. Macmillan] to say that he was rejecting their application for making the new town in Mobberley and Lymm. Lady S thought that as a result of this it was almost certain that the question of Congleton would have to be investigated again. Lord S was also extremely gloomy about the whole situation.

I wrote to D.S.I.R. to acquaint them with the potential dangers ahead: '*Nov 15* . . . reassuring letter from Greenall about the new town question. He had received every sympathy in his dealings with the Ministry who had assured him that in the event of any proposal coming to them they would ascertain that it would not interfere with the performance of the telescope.'

The next move was not long in coming. Early in 1955 a series of events began which quickly led to a major confrontation.

Febr 4 [1955] . . . there have been further alarms about the building up of Congleton . . . concerned at the weekend when the local Congleton Chronicle contained an announcement that the Chester [county] and Congleton borough authorities were to meet the Manchester representatives on Tuesday to discuss new plans for rehousing Manchester's overspill in Congleton. I phoned the Town Clerk [of Manchester] on Monday morning to remind him of the letter which he had in his file from me about the influence on the telescope of rehousing Manchester in Congleton. He assured me that Manchester had no interest whatsoever in building up in Congleton and this was entirely an approach due to the Chester people . . . the report in the paper on Wednesday morning about this was very unsatisfactory from our point of view. The result appeared to be that Manchester would consider sending out 50,000 people to Congleton if the Board of Trade thought it possible to collect sufficient industry. I sent the cutting of this to the D.S.I.R.

There was a lull. '*Feb 21* . . . the excitement about overspill in Congleton has died down and the local paper this weekend came up with a great moan that Manchester was not interested in Congleton, and in fact the Clerk to the Council said that his view was that Manchester was not interested in over-spilling anywhere.'

The quiescent period in local affairs continued until May. In that month the local weekly *Congleton Chronicle* started its campaign against Jodrell Bank and the University. By avenues which I never discovered the paper had got information about the memorandum on restriction of development in the vicinity of the telescope which I had sent to the Ministry of Housing through the D.S.I.R. The paper described it as 'a bombshell which is putting a ban on development, not only of Congleton, but the whole of the surrounding district'.

At first I ignored the campaign. Then, suddenly, in the third week of the month I had phone calls from Hingston to say 'that the Congleton issue had assumed more alarming proportions because it was now being taken up by the national press', and calls from Greenall and Cawley asking me to come to an urgent meeting at the Ministry to discuss the Congleton situation.

May 25 [1955] . . . yesterday afternoon . . . meeting at the Ministry of Housing with Greenall and Smith-Rose who the Ministry wanted in order to get the official view of a government scientist. I had already received the detailed plan, via the D.S.I.R., of the Congleton proposals. The ones which really concerned us were not those concerning slum clearance which were outside the 6 mile circle but the West Heath development . . . we raised the strongest objections to these and in this I was wholly supported by Smith-Rose. The difficulty of resiting industry and housing is, apparently, that Congleton would be involved in the construction of a new sewer costing £115,000 . . . Under great pressure and with considerable reluctance we finally agreed that the Ministry should inform Congleton that they would allow housing to take place only in the southern section of this West Heath area south of the road to Newcastle but that we raise no objection to the building of the proposed County College and Schools in the northern part of this area. With regard to the industrial area, we were insistent that this would have to be resited.

A near neighbour and friend in the village in which I live was an industrialist—Mr. L. S. Hargreaves, the founder and director of Aerialite Ltd.—and, unfortunately, on the same day that I received the maps of Congleton's proposals from the Ministry I had the documents from the local planning authority indicating that he had bought twenty acres in the proposed Congleton industrial zone to develop his factories. I had no real alternative but to raise strong objections to the proposal: 'Laurie Hargreaves has already written to the local paper saying that he will fight this ban since he has already invested £100,000 on the purchase of an existing factory and land in this area.'

A week later I met Laurie Hargreaves at a parish meeting in the local village school. He told me that the Member was going to raise the issue when Parliament reassembled. Things rapidly got worse.

June 3. This morning there was a telephone call from the Manchester Evening News asking me if I had seen the lengthy memorandum which had been circulated to all newspapers and local councils by the Cheshire Federation of Ratepayers. . . . this memorandum apparently pointed out the acute dangers facing the district because D.S.I.R. were seeking to enforce building restrictions in the neighbourhood of the radio telescope . . . went to extremes by suggesting that we were intending to prevent the erection of any houses or even to prevent people using television receivers in this 6 mile radius.

On 6 June I had my first real indication of how serious the situation was. A friend had asked me to talk to the local Congleton Rotary Club, and in the hope of making an influential group of the local people fully informed I had agreed to do this. In the evening (a Monday night), when I arrived in Congleton, I found that instead of a dozen or so people as I had expected, the meeting had been transferred to the large hall of Danesford School which was packed to the doors. The assembly was hostile in the extreme, clearly interested only in the problem of local development. Question time was the opportunity for long speeches from the floor. Laurie Hargreaves was in the front row. He began by asking me 'What would you do if you had spent £100,000 on an assurance from the local council that you could develop a factory site and then be informed that permission could not be obtained because of the radio telescope?' Since the 'local council' were there in force the appropriate answer could hardly be given.

One incident I still recall vividly was the speech of a councillor who rose, red with rage, and pointing his finger at me shouted 'You have sterilized the whole of the West Heath sewerage scheme—why don't you go to the Sahara?' The next issue of the local *Chronicle* was devoted almost entirely to the radio telescope and the development of Congleton. I wrote:

> The most astonishing thing about the whole of this business is the, apparently, genuine belief by the Congleton Councillors that our objections to their development plans originated only a few weeks ago. They will not accept my statement that our objections to Congleton's westward developments have been made over periods of years.

Events had passed the danger point. The attack was becoming personal. My mail was sprinkled with angry letters from local people. In the shops the assistants withdrew as from a leper, whispering behind their showcases 'That's him'. My disinclination for shopping became absolute. As the attack developed, fanned by the personal abuse in the local paper, I withdrew entirely from the Congleton scene and although my home is only four miles from the town many years passed before I walked in its streets or shops again.

I have explained already the reluctance of the University to be associated openly with these local problems. In the person of Rainford the University was, of course, fully acquainted with the developments but the daily battles were being carried on by me as a lone individual. I decided that, at least in the case of Congleton, I needed support. I asked Rainford to come out so that I could give him a detailed story of these Congleton developments. He did so on 17 June and I gave him copies of all the appropriate documents. It is as well that I did so: 'Laurie Hargreaves came to see me . . . he intends to sue the Crown for a quarter of a million pounds since he has in writing, authority to carry out these developments and he is seeing the Minister on Wednesday.' In the local paper that weekend it was stated that the local M.P., Air Commodore Harvey, was to ask two questions in the House on Tuesday. Firstly, what restrictions the Minister had placed on building developments in the neighbourhood of Jodrell Bank and secondly, if the Minister would now give immediate authority for Congleton to proceed with its West Heath sewerage scheme. The answer given to these questions was, in effect, that there would be a public inquiry over the objections which we had raised.

The Town Clerk of Congleton asked if he could visit me and did so on 28 June. In these conversations, based on advice which Rainford had given me, I suggested a compromise which would allow Laurie Hargreaves to use his factory and also some limited building to take place in one of the West Heath areas. On their part Congleton would withdraw claims for further industrial development on that site and would withdraw claims to develop West Heath inside our six-mile zone. '. . . surprising that Congleton is willing to proceed with this modified scheme [I wrote on 28 June] but somehow it all appears to be connected with their sewerage scheme . . . on the whole, therefore, there are grounds for hoping that the tension over Congleton will decrease, and the agonising interlude of the public inquiry may not, after all, have to be faced.'

July 6 . . . to my amazement I have been told that the Congleton Council have rejected the offer of compromise and that the Ministry have proposed to call a meeting to discuss the situation with all concerned. . . . also shocking disclosures in the local paper last week on what Laurie Hargreaves is supposed to have said to me in my office . . . wrote violent objections to him and to the Town Clerk, and I understand that the editor has promised to make a withdrawal in next week's issue.

Air Commodore Harvey proceeded to press the Congleton issue in the House and asked more questions on 12 July as a result of which the Parliamentary Secretary to the Ministry of Housing convened a meeting on 26 July of the parties involved. Harvey led the Congleton delegation of the Town Clerk, Surveyor, and two Councillors. Cheshire County Council was represented by the Clerk, the Planning Officer, and other officials. The D.S.I.R. were represented and it was with no feeling of pleasure or anticipation of an easy passage that Rainford and I travelled to face this assembly on that hot July day.

We need not have worried. I was tremendously impressed by the way in which the Parliamentary Secretary[1] dealt with the problem. He refused to discuss past history or issues of principle. When Harvey suggested that the telescope should be moved 'he became very fierce and said that the Government would not consider the possibility for one moment and that it was a waste of time to discuss it'. When the Congleton deputation began to abuse me personally he dealt with them in a peremptory manner.

Eventually a compromise was reached which differed extremely little from the proposals I made to the Town Clerk on 28 June. The Minister drew a line on the map which became the famous 'F line', prohibiting development within our six-mile zone. The Congleton Council had rejected my proposals; now they accepted almost the same ones from the Minister with joy: the next issue of the local paper hailed it as a 'victory for Congleton'. 'Rainford and I felt quite happy after this meeting, not only because it seemed that the particular Congleton issue might at last be settled but also because of the impression of solid

[1] W. F. Deedes, the Conservative M.P. for Ashford (Kent).

official backing which was apparent.' The second assessment was correct, the impression about Congleton was, alas, quite wrong, although the recurrence of the trouble belongs to a later period. On 22 August I wrote, 'The Congleton issue is quiescent and for the first time for months the local paper this week contained no mention of Jodrell Bank.'

Years later the Congleton Council renewed its attack on the F line restriction and it remains today an anxious issue. These and the many other planning troubles in the neighbourhood have constantly recurred but lie outside the epoch of this part of the telescope story. It was fortunate that, following the earlier ministerial advice, we were able to develop a consistent policy with the Cheshire planning authority. Clearly we could not ask for an absolute prohibition on any additional building within miles of Jodrell Bank. The solution was to agree with the planning authority that certain regions of existing small villages and townships should be the subject of infilling and minor extension. Outside these regions, particularly in a zone extending to four miles south of the telescope, we would object to building unless it was for agricultural use or there were other compelling reasons. These arrangements have worked well, because we have been at one with the county planners in opposing attempted infringement of this policy. Frequently the appellant has gone to appeal and on these occasions—sometimes for industrial development and often for private houses—I have had to appear at the inevitable public inquiry.

It is a sad commentary on our age that most of the private individuals who have gone to appeal have done so because they saw the chance of making a considerable sum of money by merely selling land for building sites, not because they wished to live there themselves. The appellant's solicitor has on several occasions asked me if I did not think the University should pay compensation to his client, a question which I am glad to say has always been rejected instantly by the Inspector as being inappropriate to the business of the inquiry.

Most of these cases have been in the Rural District of Congleton and the inquiries have been in their offices in the small town of Sandbach. On the whole they have been uneventful, and the final recommendations have not so far been in opposition to our needs. Two occasions, however, remain in my memory. It was my practice to ask my assistant, Mr. R. G. Lascelles, to attend for the whole inquiry so that my own intervention could be limited to the time at which the University's case was to be presented and the subsequent questioning. At one inquiry I arrived, as arranged, in the early afternoon when the business had already been proceeding for the entire morning. To my dismay I recognized the Minister's representative who was conducting the inquiry—a cousin whom I had not seen for over a decade. I felt it prudent to declare my relationship, but the appellant graciously allowed the proceedings to continue. The second occasion echoed around the world. I was at that time in the middle of some night observations with the telescope and for several nights previous to the inquiry I had been infuriated by a source of interference which occurred with surprising regularity at 11 p.m. In the loudspeaker I easily recognized the source of interference as the ignition from the type of engine used on mopeds. However, the source lasted for ten to twenty minutes, and a moped travelling along one of the nearby roads would be recorded for a minute or less. Furthermore, the rate of the individual noise sources varied, precisely as would occur if the driver were stationary in the beam of the telescope, occasionally opening his throttle and then closing it again. In order to make my point to the Inspector I displayed one of my records at the inquiry and remarked that the effect was as though a young man had taken home his girl friend on the moped, but was reluctant to say goodnight. At that time of night the telescope was investigating a star in the direction of Holmes Chapel. A sleepy local newspaper reporter suddenly awoke, rushed out of the room, and by the morning most newspapers in the world had the story of my 'eavesdropping' on the courting couple.

Electrical troubles

When I first came to Jodrell Bank there were no public electricity supply mains within miles of the place. The villagers used oil lamps or candles and a few of the larger country houses had a private electricity supply plant. For radio astronomy this was good. Overhead mains generate electrical noise and the higher the voltage the worse it becomes. Electrical gadgets used in and around houses often spark and radiate more energy into a radio telescope than an entire extragalactic nebula.

We soon had to resign ourselves to the fact that this ideal situation was unlikely to last. It would have been unreasonable and unrealistic to have opposed the bringing in of low-voltage lines to feed the surrounding villages. However, I reacted sharply early in 1952 when I learnt of the proposals to erect the Drakelow–Carrington 275 kilovolt line for the 'supergrid' scheme. On the suggested route it would have come within 7,000 yards of the telescope.

Objections were made to the Electricity Authority and as a result a meeting was arranged at Jodrell Bank on 10 July 1952. Officials responsible for the route of the line were accompanied by a member of their research division who had recently left Jodrell Bank to join the

Authority. He gave us details of the magnitude of the noise emitted by a line at this voltage on various frequencies from which we calculated the effect on the telescope.

The result of this calculation was extremely depressing. With the full gain of the paraboloid looking in the direction of the line it appeared that the interference would be a million times worse than the present noise level. Even with good conditions with the paraboloid not looking in the direction of the line it appeared that one still had to contemplate a tenfold increase in the receiver noise level.

There was an endless discussion which reached an impasse and the only practical suggestion made at that meeting was that the Electricity people would look into the possibility of switching on and off a 132 kV line which ran six miles west of the telescope so that we could make our own measurements on the effect of this line. I noted that 'the meeting was of an extremely irritating character not only because of the depressing information tabled but also because of the unhelpful attitude of the three visitors'. Indeed this initial confrontation with the Authority was, unfortunately, to prove a prototype of much that was to follow in the years ahead.

The next day (11 July): 'further discussion with Hanbury Brown and Davies about yesterday's meeting on the grid line. The gloom and despondency was confirmed and continued. It was agreed that I should write to the planning officer objecting in the very strongest terms to the erection of this line and asking that it be resited to run north east of Prestbury and east of Macclesfield.' We also made our internal arrangements for making tests on the 132 kV line.

On 24 July I received a memo from the Research Laboratories of the Authority on the 10 July meeting. I replied that the notes were correct in so far as the meeting was concerned, but I objected to the inclusion of uninformed comments about the levels of interference to be experienced from other sources in the neighbourhood with which the writer tried to justify the additional source of trouble which would be caused by this line. Again, this set a precedent of attempted self-justification, which was to persist throughout all our subsequent dealings with the Authority.

The testing on the 132 kV line took place during the night of 5/6 August 1952, and since I have been critical in general terms of our contacts with the Authority I ought to say that in these kind of detailed arrangements the local individuals have always been exceedingly helpful. The controls and communications necessary for these tests put the North Western Area people to much trouble and inconvenience but they extended themselves to meet our requests. The results of our own measurements on this line cheered us slightly as regards the effect of that particular line on the telescope, but left no doubt as to the seriousness of the 275 kV line project.

The uncompromising and unsympathetic letters from the Authority made it inevitable that a deadlock would be reached over the 275 kV line and in September 1952 we received a letter from the Ministry of Fuel and Power to say that they proposed to call a meeting of all concerned to attempt to resolve this problem. Eventually the meeting convened in the offices of the Ministry on 7 October.

The chairman of this meeting was the Minister's deputy for electrical affairs. He entered a room full of people. Surprised at the numbers before him he asked his secretary quietly who they were: 'From the electricity authority, sir.' 'Where are the University people?' 'He is here, sir.' I was outnumbered by fifteen to one. The Minister's deputy was not at all impressed by the largeness of the Authority's deputation and asked his secretary in my hearing if they had no other work to do. I do not suggest that the final judgement was influenced by this situation but at least the size of the confrontation had the psychological effect of placing me in company with the Minister's deputy as the party resisting the invasion of his office by these aggressive individuals. At the end of the day the Electricity Authority were instructed to re-route this supergrid line away from the telescope.

Although this confrontation had been resolved in our favour I remained unhappy. Yet the solution was as good as could be found under the circumstances. If we had attempted to force the line even further away it would have had to cross the hills and the increase in its height would have nullified the advantage gained by increasing the distance. We had asked for it to be placed underground but of this, so we were told, there was no hope because even an experimental section at this voltage would cost £500,000 per mile. Furthermore, it seemed to me that the Authority was developing these overhead lines with ever-increasing operational voltages and gave little attention to the consequential problem of interference and means of suppressing it. At least the amount of information which we were able to obtain from them about measurements on these matters was negligible. I wrote a complaining letter about this to D.S.I.R. and to the Scientific Adviser to the Electricity Authority who had remained aloof during these various troubles. On 28 November I had his reply suggesting that some actual measurements instead of speculation might be useful—precisely the point at issue! He 'ends his letter', I recorded, 'by saying that he does not think that I have been fair to the officers of the B.E.A. since they have certainly placed all their available information at my disposal but this was not the point which I made in my letter. I complained that they were unwilling to start up any new research on the problems.'

The 275 kV supergrid line was the first of our major issues with the Electricity Authority but others soon followed with the regional boards. In the following spring we were acquainted with a proposal to run a 33 kV line on the western side of the telescope. On 25 May 1954, the area planning office sent me a map '. . . showing a proposal for running another grid line from Crewe to Knutsford in order to feed the Salt Works of Murgatroyds near Middlewich. The projected run of this line juts right into Holmes Chapel and is completely out of the question as far as the radio telescope is concerned. I replied that this line could only be tolerated if it was erected at least as far away as the existing line between Knutsford and Crewe.' It really did seem extraordinary to us—quite apart from any affair of the telescope—that with one line of pylons running from Crewe to Knutsford, another line was planned between these two centres along an entirely different route.

Again we entered battle, this time on a golden September day far too lovely to be spent in the Liverpool dockside premises of the North Western Electricity Board.

Sept. 19, 1954 . . . On Thursday Hanbury and I spent all day at the N.W. Electricity Board in Liverpool. They produced new measurements to try to prove that the 11 kv lines already near Jodrell would give as much interference as their proposed 33 kv line. Of course we refused to budge; all that they have to do is to move the line a few miles west to the run of the 132 kv line. They are very obstinate about this on the grounds of wayleave difficulties, but we were even more obstinate. It was really a trial of endurance which continued until after 4 p.m. Finally they reluctantly agreed to look into the possibility of moving near to this line. The situation is absurd. The line is required to supply 15 megawatts to Murgatroyd's at a revenue of £1 million per annum, and yet the Electricity Authority is squabbling about the expenditure of a few thousand pounds involved in resiting this line.

Our day in Liverpool was not wasted: the line was resited!

Early in 1955 publicity was given to the plans for the electrification of British Railways. The first major enterprise was to be the Euston–Manchester line and the first test section was to be the Crewe–Manchester section, which, between the stations of Goostrey and Chelford, runs within a few hundred yards of the telescope. I could hardly believe that the news was correct: only five years previously when we examined the implications of the site of the telescope we were informed that the electrification of this section of the line was not to be contemplated for at least fifty years. Instantly I wrote to D.S.I.R. asking them to inform British Railways that it would be necessary to undertake complete screening of the line in the vicinity of the telescope. D.S.I.R. gently evaded the task by saying that they would prefer us to do these negotiations ourselves 'in the first instance'.

Fortunately it transpired that one of the University officers had the necessary communications at high level and more than a year later, on 4 June 1956, a party of railway engineers visited me at Jodrell Bank for a discussion of the problem. The seriousness of the situation was not in doubt: data produced by the engineers at that meeting led to an estimate that the interfering signal from the railway would be something like 300 times the minimum signal detectable by the telescope. 'The B.R. people seemed quite willing to consider a cage should this be necessary.' The further development of this story lies beyond the scope of this book; the lines were not erected until the telescope had been operating for several years. Fortunately the good impression created by the railway engineers at this meeting continued. There were certainly troubles, but the Chief Engineer's department and the British Railways organization exceeded all my expectations in the assistance and help which they gave in isolating and then curing the interference which arose when the trains eventually came into service. It was, perhaps, a fortunate circumstance that one of their number was an enthusiastic and knowledgeable amateur astronomer.

Radio frequencies

In addition to the troubles caused over a broad frequency band by interfering signals from power lines, electrical devices, and ignition, radio astronomers everywhere faced an even more serious problem because the radio waveband was already nearly monopolized by other users. Before the advent of V.H.F., radio listeners were only too well acquainted with the fact that the radio waveband was overfull with transmissions. V.H.F. helped because the distance over which signals could be received was much less than on the lower frequencies. Broadcasting for television and radio entertainment was merely a small part of the huge problem of accommodating everybody's interest in the restricted band of wavelengths: the armed services, the Home Office, commercial enterprises, civil aviation, and a host of other interests all laid claim to use of radio frequencies for communication purposes. In the United Kingdom frequency allocations are made by the General Post Office based on broad arrangements made at assemblies of the International Telecommunications Union for the use of various bands of frequencies by the differing users.

Radio telescopes now added to the difficulty. Although the needs of the radio astronomer were relatively small, the enormous sensitivity of the radio telescope required international clearances of a range of frequency bands covering the whole spectrum of radio frequencies. The study of the radio emissions from a distant nebula could all too easily be obliterated by the radiation from a transmitter across the Atlantic, which although out of the line of sight could, for instance, be reflected from the moon. This is not a speculative example. Unfortunately precisely this problem has arisen at Jodrell Bank when working on a frequency which has been cleared for Europe but not for the rest of the world.

The case of the radio astronomer was weak in the international forum. It was simply a case of a fundamental scientific study of the universe in competition with an avalanche of imperative demands for allocations from organizations where either big money or security and defence were involved. Martin Ryle was molested by this problem in Cambridge as much as we were at Jodrell Bank and we soon realized that only an approach at the highest possible level would be likely to gain us any sympathy. We therefore jointly worked through the Royal Society because the Society is responsible for the national representation at the international assemblies of many scientific activities. One of the Royal Society Committees was the British National Committee for Scientific Radio, which as a main activity was concerned with U.R.S.I. (the International Scientific Radio Union). We were both members of this Committee and repeatedly made complaints that unless the claims of radio astronomers to a share in the frequency allocations received serious attention, then within a few years it would be impossible to continue the researches with the radio telescopes. Eventually the Council of the Royal Society agreed to send a recommendation from its British National Committee for Scientific Radio to the Lord President of the Council and the Postmaster-General.

On 20 October 1953 we were summoned to a meeting at the G.P.O. Headquarters in London. Nearly every interested party must have been present at this meeting: the War Office, Navy, Air Force, Civil Aviation, Home Office, the Chief Engineer's department, and the D.S.I.R. and the Lord President's Office. The expressed purpose of the meeting was to assist the Lord President to draft a reply to the Royal Society's memorandum. We had tabulated our requests for radio astronomy, essentially an allocation of a few megacycles bandwidth every octave throughout the spectrum, and the meeting explored the consequences of securing these allocations. Everywhere there was trouble: there were reservations either for Civil Aviation, police, taxis, maritime services, or enormous chunks for the armed services, broadcasting and television. There were no conclusions to this meeting except that we were able to demonstrate the urgent needs of radio astronomy, and that our demands must be taken seriously. '*Nov. 3.* Ryle . . . regards the issue of the 200–400 Mc/s band as a test case . . . it is absurd to believe that the Air Ministry cannot release 2% of this for other purposes. Adds that if this is the case then their planning is too marginal and hence is not realistic.'

Inevitably at this stage whenever we caused trouble of this nature an instant reaction was that we should abandon the construction of the telescope.

Thus on 5 December 1953:

Greenall (D.S.I.R.) phoned to say that the Lord President was about to see the Postmaster-General on the question of our frequency problems. He asked how much of the work had actually been completed so that the Lord President could answer questions about moving the telescope. . . . I said that this idea which had already arisen as a means of evading the troubles with grid lines and new towns was fatuous.

Meanwhile Ryle had arranged for both of us to tackle the Air Ministry to try to persuade them that they could yield in their refusal to give up a band in their enormous allocation. This band blocked an important part of the radio astronomy spectrum and we knew from working illegally in it that the usage rate in some sections of interest to us was exceedingly low. Unfortunately I missed the meeting completely. Having caused much trouble by asking for it to be rearranged from the afternoon to the morning, the Ulster Express on which I was travelling remained stationary for two hours because of a derailment. Alas, Ryle told me later that it was a poor meeting as far as our interests were concerned.

There was a constant stream of correspondence between Ryle, myself, the G.P.O., and the D.S.I.R. on problems arising from the October meeting, and finally in late January or early February 1954 the Lord President replied to the Royal Society. The Executive Secretary of the Royal, David Martin, showed me the correspondence on 3 February. An important step had been taken. The Lord President stated that a decision had been made to appoint members of the D.S.I.R. to the various committees dealing with frequency allocations so that they could look after the interests of radio astronomy. However, the Lord President also indicated in his letter to Sir David Brunt (Physical Secretary of the Royal) that the discussions had resulted in many of the astronomers' difficulties being overcome. I was asked to comment so that Brunt could reply. 'I said that the appointment of members of the D.S.I.R. to the committees was a most important step but that in the vital frequency band of 30 to 1000 Mc/s we had achieved no success in effecting frequency recommendations whatsoever' (3 Feb.).

On 26 March the results of this Royal Society approach were discussed at the meeting of the British National Committee for Scientific Radio which had originally instigated the approach.

Both Ratcliffe[1] and I explained that we were grateful for the help which we had been given. R complained bitterly that a 79 Mc/s transmitter had been set up 15 miles from Ryle's interferometer and I complained that having shifted a frequency of one of our equipments on the advice of the Post Office, at much cost and time, we were unable to work because of transmissions from American aircraft. Group Capt. S said he was sorry to hear that we were not more pleased.

The significant part of the Royal Society approach was that we now had official representation in the hierarchy of committees responsible for allocations. Dr. R. L. Smith-Rose, who was then the director of the D.S.I.R.'s Radio Research Station, was our official representative. He showed me after the meeting correspondence from the Secretary of State to the Under Secretaries for Air, Sea, and Army which were clearly designed to be helpful to us.

It is not appropriate to continue this story in detail. We were to have endless national and international troubles with frequencies. However, by this early action we had achieved a national status for the subject and Smith-Rose has continued, even after his retirement from his official post, to represent our needs firmly throughout the tiresome and lengthy national and international meetings which have resulted.

[1] J. A. Ratcliffe, C.B., C.B.E., F.R.S., was then Reader in Physics and Fellow of Sidney Sussex College, Cambridge. He succeeded Smith-Rose as director of the Radio Research Station in 1960 where he remained until his retirement in 1966. Ratcliffe was a friend of long standing and his own distinguished contributions to ionospheric research led him to be particularly interested in the work of Ryle and ourselves. His help to us both in many matters like this was invaluable.

29

Public relations—the Press

Our initial relations with the Press were uneventful. Our official contact was the P.R.O. of the D.S.I.R., Colonel Walter Hingston, and as we discussed the initial press release in the spring of 1952 we had little idea that together and separately we were to suffer turmoil for which we were ill prepared. The release of the news of the decision to build the telescope was timed for 11 a.m. on Friday 25 April 1952. On 24 April I had my first introduction to Press leakage. The Assistant Editor of the *Manchester Guardian* phoned to express surprise that they had not been given news, in view of the local interest and association. I could not understand how he had any information at all twenty-four hours before the time of the official release. The answer came from Hingston: the eagle eye of the *Daily Telegraph* science correspondent had extracted a note from the D.S.I.R. annual report and had made a timely publication, gaining twenty-four hours on his colleagues without compromising anyone. The papers of Saturday 26 April gave the news a good reception and *The Times* carried a pleasant leader on 'a great scientific project'.

We were then left in peace and I doubt if it ever entered my head that the Press was to be a significant factor in our future history. Hingston, on the contrary, saw many possible troubles. On 23 July 1953 he came to Jodrell 'pleased to see the progress on the foundations, but said that he had become extremely worried about the publicity situation here. He thought that as soon as the steelwork became visible from the road, we might be submerged under an avalanche of sightseers and visitors.'[1]

Occasionally we appeared in the newspapers: for example, over the

[1] It is interesting that even at such an early stage Hingston was worried about the visitor problem. Hanbury Brown said that Palomar attracted 150,000 visitors a year. I did not believe this but Hingston wrote to Dr. I. S. Bowen the director of the Mt. Wilson and Palomar Observatories who confirmed that this was the case and explained how they were dealt with. Hingston then produced a paper for the University recommending that permanent arrangements for visitors should be put in hand. After 13 years of discussion this was done and the Concourse Building for visitors was opened at Jodrell Bank on 7 May 1966.

case of Col. Erroll's question in the House about town development described in the previous chapter. My first information about this came from the *Manchester Evening Chronicle* on 10 November 1953, and that same evening I agreed with a *Manchester Guardian* reporter the wording of a paragraph which he would publish in the next day's issue. What actually appeared was a travesty of the agreed statement. I was so incensed, first of all by Erroll's question and then by this failure of the *Guardian* to publish the statement which I gave to their reporter, that I complained to the Editor, A. P. Wadsworth. For some years before the war I had been A.P.'s neighbour and many years of friendship enabled him to deliver to me a sharp reprimand in one of his characteristically short letters, for which I still remain grateful.

The doleful occasion of our first massive Press visit to Jodrell Bank on 10 June 1954, has already been mentioned in Chapter 15. The concept of this visit was Hingston's. He was worried about the potentially unfavourable effects of the reports of the first investigation of the P.A.C. on the telescope finances. In March of that year he had asked the science correspondent of the *Manchester Guardian* to publish a piece about the telescope a few days before the first P.A.C. inquiry on 25 March, and he was anxious to create a favourable reaction amongst the Press as an investment against the gloomy future which he was already visualizing. Unfortunately, as I have described on page 70, the good intentions misfired in a midsummer quagmire.

Hingston immediately made plans to redeem this misfortune as soon as the steelwork became impressive. The next effort was in March 1955 when there really was something of the telescope structure to be seen. This was a major event with many of the picture papers coming days in advance. The main Press day was on Tuesday 22 March and it seemed to me that everything had gone well. To judge from the Wednesday morning papers the reporters had also been quite impressed and there were many favourable accounts of progress and of the scientific significance of the project. For me, however, the pleasure was short-lived. Before twenty-four hours had elapsed the whole event turned into one of the sourest of memories. Travelling in the train to London on that Wednesday morning I was oblivious of a telegram which was on its way to me. On that afternoon I gave the Trueman Wood Lecture at the Royal Society of Arts. As I was walking into the lecture room behind the Chairman I was handed a telegram. It was from Husband:

> We are very concerned at lack of acknowledgement in Times and Picture Post. Picture Post wording most unfortunate. Will you please write Times and Picture Post that developments and design by Consulting Engineers Husband & Co. This matter considered most serious and urgent.

This was followed by a second telegram handed to me after the lecture: 'Please add Daily Telegraph and Manchester Guardian to my earlier telegram.'

I was bewildered by this. The Press had obviously been more interested in the astronomy than the state of the engineering as it was at that stage and their published accounts presumably reflected this interest. As soon as I could after the lecture I phoned Hingston at D.S.I.R. who most strongly advised me to do nothing whatsoever in regard to writing to the papers. The correspondence which followed between me and Husband and Hingston is amongst the most acrid of the thousands of communications which were exchanged during those years. This was the immediate prelude to the meetings with Husband in early April (described in Chapter 21) which revealed the enormous debt which we were incurring.

Only in the calmer atmosphere of later years was I able to appreciate Husband's point of view. He felt quite rightly that unless the engineers, not only in his office but also in the factories, who were extending themselves to do this extraordinarily difficult job, were given their due share of acknowledgement, then we would lose their goodwill. It was indeed an important point. Everybody on the job could have earned their profits commercially far more easily by conventional activities. Although I was the scientist on the job and not the engineer I did my best to make sure that in the future there could be no complaints of this nature and as far as I can recall there never were, either on this telescope or those that followed it.

In the light of my experience with the Press in later years I find it strange that the huge debt which we were incurring on the telescope escaped their scrutiny for so long. In fact a year elapsed after our own realization that we were at least a quarter of a million pounds in debt (Chapter 21), before we were troubled by the Press on this matter. Then on 13 April 1956: 'A Sunday Times man phoned this morning about some new expenditure on "foundations". I told my Secretary to put him on to the Bursar.' Later in the day John Maddox[1] (then Science Correspondent of the *Guardian*) asked me if I knew that 'a newspaper—which he admitted to be the Sunday Times, was coming out with a story about changes involving £270,000'. The *Sunday Times* did indeed come out with this story (on 15 April). We were greatly alarmed, but the story died, or at least was not followed up in any major way 'accord-

[1] Until 1955 Maddox had been a Lecturer in Theoretical Physics in the University of Manchester. His professional association and friendship with us was to turn out to be of great help and value subsequently during his career as the *Guardian* Science Correspondent (1955–64). In 1966 he became Editor of *Nature*.

ing to Rainford, because Hingston handled the Press very well about it'. Nevertheless, we were uneasy and Hingston's discovery of the source of the leakage increased our disquiet.

In my anxiety to keep the Press from news of our troubles throughout those dreadful days I made a frightful mistake in August of that year. Still smarting from the consequences of the 1955 Press visit and terrified of publicity on the financial front while so many delicate matters were at stake, I did everything in my power to keep information from the Press. The pursuance of this policy led me to the cardinal and stupid error of August 1956. The British Association was meeting in Sheffield and we had agreed with the Secretary that Jodrell Bank could be included in the itinerary of visits for the scientific sections. A short time before the meeting I received a further request that members of the Press should also be included in the visit. I informed the Secretary that on no account could the Press be admitted, and since I had to visit Sweden for a scientific meeting on the date of the visit I also left firm instructions at Jodrell Bank that no members of the Press could be allowed inside the establishment.

Then, on Monday 3 September: 'Arrived back from Sweden this morning to find the Press in a turmoil over the telescope because I had refused to allow them to come with the B.A. visit last Saturday.' *The Times* had a snooty article about the instrument. It was front-page news in the *Manchester Guardian* and the *Daily Express* announced that I was to be carpeted and asked to explain the reasons for my ban 'to a committee of the Treasury and Press Association'. Apparently telegrams had been sent to the Vice-Chancellor and me the day before the visit but I was out of the country and knew nothing of this incipient trouble.

I was especially bothered about *The Times* article and sought elucidation from the Editor:

. . . your paper of Sept. 3 contained an article on the radio telescope at this establishment, from which it appears that your correspondent obtained access to the Station on Sunday Sept. 2. The entrances to this establishment carry prominent notices which forbid the entry of unauthorized persons and I should be glad if you could inform me from whom your correspondent obtained the necessary authority, and from whom he obtained the misleading information about the telescope printed in this article.

My complaint was passed on to the Deputy News Editor who replied:

I must point out that our correspondent was acting in fulfilment of his prime duty, which is to seek news wherever it is to be found. In this we would wish to uphold him to the full. I am not prepared to divulge the source of the authority which he obtained. . . . With regard to your complaint that we obtained 'misleading information' about the telescope, may I say that we have received no complaints from any sources including the D.S.I.R.

It was obvious, although I had no actual proof of this, that *The Times* correspondent had, in fact, infiltrated on the Sunday afternoon, evading the man whom we had guarding the entrance, and gathered his information from the contractor's men who were at work at that time. No member of my scientific or other staff saw this correspondent. For these reasons I was most dissatisfied with this letter, and wrote again on an informal basis to the Deputy News Editor in an effort to find out what led their correspondent to behave in this highly irregular manner. I had a polite reply but no satisfaction. 'Our correspondent decided to present himself at the site and there he gathered his material by means which are his own concern. He is now out of the country and I have nothing more to add on this aspect of the matter.'

If this had been an isolated example no one would have worried. Alas we were in no position to bear the antagonism of the Press, and I was to pay dearly for my over-zealousness in trying to keep the reporters away from the load of dynamite on which we were sitting. I have already referred in Chapter 26 to the publication of Sir Frank Tribe's report on 5 March 1957, and this presented the golden opportunity for a major onslaught by the Press.

'*March 6* . . . Yesterday Sir Frank Tribe published his report and among other things he spilled the whole story about the financial chaos on the telescope. The newspapers are having a field day what with this and the theft last week.'[1] Chapter 26 describes why the timing of this publication was so unfortunate and Husband's understandable annoyance at the biased publicity.

Hingston, who knew the pulse of the Press by instinct, was beside himself with anxiety. He told me that various parties in the Press were against me and had threatened to retaliate because of the incidents last September on the occasion of the British Association visit. He was afraid that they were reserving their full blast for the publication of the report of the Public Accounts Committee. '*April 18 1957*. Our one hope is that the report may be delayed until July or August in which case the telescope may be finished. Hingston is even toying with the idea of

[1] One of our outstations had been ransacked and some electronic equipment removed. This had nothing to do with the telescope but the Press seemed to have their eyes on us whenever we moved. Someone reported the incident and even *The Times* phoned me at midnight to inquire if I suspected the activities of an international agency!

convening a Press conference to deal solely with the scientific aspects of the telescope so that I could imbue them with the idea that £100,000 or so was trifling compared with the scientific importance of the instrument.'

The 'specialist' Press conference never materialized. Weeks elapsed and mercifully the P.A.C. report did not appear as soon as we feared. As the telescope moved to completion there were insistent demands for another Press conference both from within and without. Husband, particularly, repeatedly said that the only way to get the multitude of minor jobs cleared up quickly was to set a date either for an opening ceremony or for a final conference when the telescope could be demonstrated. A formal opening was out of the question because of the aura of disgrace which surrounded the enterprise. In late May and early June 1957 the Vice-Chancellor and Hingston decided that a full-scale Press conference could no longer be delayed. Husband said that the telescope could be demonstrated in motion at the end of June if the contractors were given the incentive, and hence the entire last week of June was allocated to this conference.

In that month, as described in Chapter 26, we were becoming acquainted with the Treasury's attitude to the clearance of the debt on the telescope. At the same time, on 18 June, 'Hingston here with further details about the Press conference, which is now fantastic. Next Tuesday 40 photographers, 3 TV news reel and BBC are coming.'

The Press conference really was a major affair. Hingston assured us that we must produce refreshments and beer and so a large marquee duly appeared. Vast quantities of literature were also produced by ourselves and by the various contractors who had worked on the project. This time, as well as providing the scientific information, we made sure that anyone who wished to publish the number of rivets in the structure, who made them, and who put them in had the information close at hand. However I lived in terror of the occasion—there were so many questions which I might be asked and to which I could not give the correct answer because of compromising the delicacy of the background negotiations.

July 1, 1957. Last week's agony is over! TV, News Reels, B.B.C., and innumerable photographers on Tuesday, Press on Wednesday and Technical Press on Thursday. Actually in the end all went rather smoothly with less disturbance than I expected—thanks to the tremendous efforts of Hingston and his staff. The telescope was moved each day and performed superbly. The only real trouble was with X of The Times who got entangled with the United Steel people and missed the essential part of the proceedings. Hingston was furious, but there was not a word in The Times—neither was there about the opening of the Radio Research Board.[1] Hingston is intending to see the Editor. The Manchester Guardian did very well with a 3 page supplement by Maddox and Co.

The eventual publication of the P.A.C. report on 13 August 1957 has been mentioned in Chapter 26. The papers were full of the business. The report and the accompanying minutes of the evidence were in some parts such a travesty of the actual situation that one could hardly blame the news media for seizing their opportunity to retaliate for any imagined or real injustices which they might have suffered at our hands. For example the P.A.C. report contained the following: 'Although the University professor, who was primarily responsible for the outline design, lived on the spot, he was not consulted so far as the Department were aware.'[2] I neither lived 'on the spot' nor in 'University apartments',[3] and the accusation of 'no consultation' was a strange interpretation of the exchange of thousands of communications between me and Husband.

This great scientific endeavour had been reduced to the level of a charade by individuals who in the rightful performance of their public duty had not taken the trouble to see for themselves what was involved, or to check from those who really knew the facts that the information which was to be published as a privileged document was correct. Here is a typical extract from the dialogue with Melville answering.[4]

Mr. Arbuthnot. Will that not be a valuable asset to him?—The number of such telescopes likely to be made is extremely small. There never will be another one of this magnitude in this country.
Mr. Hoy. It does not require very many at this price.
Mr. Arbuthnot. They are also extremely large?—Yes.
So, therefore, each one is particularly valuable is it not?—Yes.

In these pages there was material enough for anything from *Punch* to a *Times* leader and not a single opportunity was lost to heap the supposed disgrace on our defenceless heads. The University of Manchester, the Department of Scientific and Industrial Research, Husband, and I, had carried the burden through nearly ten years of this feat of British science and engineering. At the moment of its initiation we received this merciless castigation. Those who were so quick to publicize our discomfiture over a matter of £100,000 or so should have turned two pages of the Report where they could have read: 'in the

[1] The new buildings of the Radio Research Station at Slough had just been opened; and Hingston was also responsible for those Press arrangements.

[2] The Committee of Public Accounts Session 1956–57, Class IX, vol. 8, para. 62, page xliii.

[3] ibid., p. 169, para. 1849. [4] ibid., p. 173, paras. 1903–5.

three years 1953/54 to 1955/56 works so started by the Army and Royal Air Force without express Parliamentary sanction were estimated to cost £4·62 million and £26·2 million respectively'.[1]

How gratified I was to the faithful John Maddox who 'came out late on Wednesday [the day after the publication of the report] and got some photographs of our first records with a nice piece about the performance of the telescope in Thursday's Manchester Guardian'.

[1] ibid., para. 70, p. xlv.

30

The Sputnik—October 1957

THE thrill of the first movements and the reception of radio signals from space had vanished in the gloom of the attacks by the Press and the P.A.C. As October dawned our affairs were in such a tangle that all seemed lost. No one likes to be associated with unfavourable notoriety. Contractors least of all; and at the end of Chapter 27 I indicated that a multitude of relatively minor jobs were simply hanging fire. The lifts broke down and repeatedly stranded us high up amongst the steelwork, we had no means of driving the telescope except from the Ward Leonard room, we had few cables, a girder had twisted in the bowl framework and we could not work at low elevations, there was no clamping device, and no one seemed imbued with any desire to get on and complete the job. For perfectly understandable reasons the relations between all parties were strained to breaking point.

Husband, in his own defence, had informed Rainford that if I was not allowed to correct the record of the P.A.C. report published in August then he 'would drag the information out of me in the witness box'. Rainford and the University Officers were in a frightful dilemma. Faced with the task of finding another £350,000 they had been presented with conditions by the Treasury (Chapter 26) which apparently made this impossible of achievement unless they either rejected me or claimed damages from Husband. They stood firm and refused to do either. For me personally there was no comfort; a great hulk of steel over which we had little control and of no effective use for scientific research, and responsibility for an overwhelming debt which had generated consequences out of all proportion to its magnitude. Those responsible for the ultimate direction of the University, particularly the Vice-Chancellor, the Chairman of Council, Lord Simon, the Treasurer, Sir Raymond Streat, and the Bursar, Rainford, never wavered in their support of me or in their belief in the telescope. The same could not be said for many of my academic colleagues who fancied that my recklessness would end up by depriving them of their freedom. The Staff Room of the University and its environs became a most uncomfortable place for me; an onlooker might have thought that I was carrying some terrible infectious disease.

The Press continued its sniping. Judging from the way in which even our minor troubles always appeared in the gossip columns we were surrounded by people with binoculars waiting anxiously to be first with the news of the ultimate collapse of the steelwork.

We needed a miracle, it seemed, to raise us out of this bottomless pit of troubles. The miracle came—from behind the Iron Curtain in the form of Sputnik I, the world's first artificial earth satellite.

The world and the Sputnik

The impact throughout the world of the Soviet success in launching man's first artificial earth satellite on 4 October 1957, will live long in the memory of this generation. The telephone bells which roused me from sleep in the early hours of the morning carried a message which I had expected; nevertheless I had no conception of the impact which this historic feat would have on the world or of its consequences for me personally and the telescope.

Even at this stage it is hard to understand why the Soviet success created such a furore of interest and excitement all over the world. The basis of the surprise was undoubtedly the lack of recognition, particularly in America, of the advanced state of Soviet science and technology. People everywhere either had not noticed, or worse, had known but failed to recognize as realistic, the portents. The excuses for this attitude were hard to sustain, particularly since plans for an American earth satellite were well known and hence the technical requirements for launching were well understood in the western world.

Precisely three years to the day before Sputnik I (on 4 October 1954), the Special Committee for the International Geophysical Year (I.G.Y.) meeting in Rome recommended that, as part of the world-wide programmes for the I.G.Y. in 1957/58, 'thought be given to the launching of small satellite vehicles, to their scientific instrumentation and to the new problems associated with satellite experiments, such as power supply, telemetry and orientation'. On 29 July 1955 President Eisenhower formally announced to the world that the United States would launch a satellite during the I.G.Y.

At that stage no one could accuse the Soviet Union of lacking in frankness about its space programme. In 1955 (15 April) the Soviet Academy of Sciences announced that it had set up a Permanent Commission of interplanetary communications, whose work included the development of meteorological satellites. On 30 July, one day after President Eisenhower's announcement, the U.S.S.R. announced that it would also launch a satellite for the I.G.Y. A few weeks later, at the International Astronautical Congress in Copenhagen, Professor Sedov, the Chairman of the U.S.S.R. Commission, said that Russia would launch a satellite within 'the next two years' and he also said that it would be heavier than the satellite proposed by the Americans.

Little notice was taken of these Soviet pronouncements. Indeed, the possibility that their technology was sufficiently advanced to make their claims feasible was widely discounted. On the other hand, the U.S. plans to launch a small Vanguard satellite were well publicized—including the setbacks in the programme—and those who were interested elsewhere in the world rested on the comfortable assumption that the world's first earth satellite would be American. The widespread attitude about the relevant status of Soviet and American technology was all the more surprising in view of the official statements which came from highly placed Soviet scientists. In June 1957 the President of the Soviet Academy, Academician A. N. Nesmeyanov, said that both the carrier vehicle and the instrumentation for the first Soviet satellite were ready, and that the first launching would occur 'within a few months'. In the same month the I.G.Y. committees were officially informed by the Soviet Union that a satellite would be launched 'within months'. On 18 September Radio Moscow said it would be launched soon, and on 1 October the Russians broadcast the frequencies on which their satellite would transmit. Three days later to the amazement of the world at large the Sputnik, weighing 184 pounds, was placed in an orbit ranging in altitude from 140 to 560 miles.

The news of the Soviet success swept the world as no other scientific and technical achievement has ever done. Sputnik I was far heavier and larger than the proposed American Vanguard satellite. It was the epitomization in a form for all to witness of the dramatic advance of Soviet science and technology in the years since the end of World War II. In particular, the scorn and disbelief which had commonly been poured on Soviet pronouncements by America almost instantly turned into a total re-evaluation of the United States effort.

On 7 October the *Manchester Guardian* carried the following comment from Alistair Cooke:

> The White House assurance, given out by Mr. Hagerty, the President's press secretary, that it did not come as a surprise, is drowned by the astonishment of the scientists, the defensive surprise of the Pentagon, and the angry cries of the Democrats. Senators Morse of Oregon, Jackson of Washington, and Symington of Missouri, were quick to blame the Administration's defence economies for the shame of the Soviets' palpable 'superiority in the long-range missiles field'. And Senator Symington applied the classic Amer-

ican poultice to every wound in the body politic. He 'demanded' a Senate investigation.

On the day of the launching a highly placed American militarist said 'anyone can fling a lump of old iron into space'. But Sputnik I was no lump of old iron, neither was it any lucky chance. If any further proof were needed the Soviets supplied it a few weeks later on 2 November by launching Sputnik II weighing 1,120 pounds, carrying the dog Laika with complete life support equipment. The failure of the Americans to launch their small Vanguard enhanced the demonstration of this relative status of achievement between East and West.

The disquiet in the inner circles in the West had deep foundations. The carrier rockets with which the Russians launched their Sputniks were Intercontinental Ballistic Missiles (I.C.B.M.s), and those who preferred to believe in the military inferiority of the Soviet Union were no longer able to do so. The immediate post-war military superiority of the United States had been short-lived: the Soviet Union had exploded an atomic bomb in 1949 and a hydrogen bomb in August 1953. Quite soon intelligence sources produced the evidence that the Soviet Union was hard at work on a ballistic missile programme. Naturally the U.S. responded with similar vigorous developments of ballistic missiles, but they did so separately in all three armed services. Furthermore, in July 1955, when the U.S. announced the intention to launch a satellite during the I.G.Y., it was decided to base the carrier rocket on the sounding rocket technology and not on the ballistic weapon development. This cardinal error and the dispersal of the U.S. efforts amongst the three services contrasted sharply with the single-minded concentrated effort in the Soviet Union. The U.S. did not succeed in launching a satellite until it adapted the Army's ballistic rocket Juno I (Jupiter C) developed by Werner von Braun's team. By that time (31 January 1958) a beginning had been made with the reorganization of American science and technology which soon led to the creation of the National Aeronautics and Space Administration (N.A.S.A.). The disarray caused by Sputnik I in America did not last long, although many years had to elapse before the N.A.S.A. and the military departments could parallel the achievements of the Soviet Union in space.

The telescope and the Sputnik

In the original 'blue book' of the telescope (Chapter 8), I inserted a chapter outlining the proposed research uses of the instrument. Amongst these was its use in the radar mode for the study of radio echoes from the moon and the nearer planets. As the ideas for the I.G.Y. national effort developed it was evident that many of the programmes proposed for the telescope would be appropriate. In particular, after the announcement in 1955 of the intention to launch earth satellites I included the tracking of these as a useful experiment using the lunar and planetary radar. We were already obtaining radar echoes from the moon with smaller instruments and some members of my staff had developed this technique for the measurement of the total electron content in the earth–moon space. These measurements would be particularly important for the I.G.Y. and it seemed to me that if we could use an earth satellite as an artificial moon close to earth then by comparison with the lunar results we ought to be able to measure the electron density in the interplanetary space between the satellite and the moon. It was an innocent suggestion made with little realization of the appalling difficulties of carrying it out—or of the quite unforeseen consequences. The important point in the history of those days is that there was an official proposal as part of the Jodrell Bank I.G.Y. effort to carry out this experiment using the telescope as a radar device to follow an earth satellite. I regarded it as a rather minor and speculative part of our I.G.Y. programme.

Naturally the I.G.Y. programmes formed a major part of the discussions during the August 1957 meeting of U.R.S.I. (the International Scientific Radio Union) at Boulder, Colorado. During this assembly the Director of the Vanguard project, Dr. John Hagen, gave a lecture on this American earth satellite. The large lecture room was packed. In his talk Hagen explained that the programme had been delayed and that a launching could not be attempted immediately. Hagen was a member of our Radio Astronomy Commission (V) of U.R.S.I. and I had known him for many years. After the lecture I asked Hagen privately what his hopes for the launching were and reminded him that we were an interested party because of our proposal to use the telescope on it. He then told me that it would be impossible to attempt a launch for several months. 'Then', I said, 'you will certainly be beaten by the Russians.' To which he replied that he did not believe there was the slightest chance of this: the Russians were known to be encountering severe difficulties and were attending a conference in the U.S. in early October to discuss them.

I was much puzzled by these opinions but could not argue further with Hagen since he was then tremendously busy with Vanguard and left immediately for Washington. In fact, a scientific friend who had spent some time in Moscow during that summer had told me only a few days before I left England for Boulder that the Soviet Union was 'nearly ready' to launch a satellite, and a news item in an American paper the day before Hagen's lecture stated that the Soviet Union had successfully tested an I.C.B.M. The news of the I.C.B.M. test seemed to me to be a confirmation of my informant's message. I did not doubt the implications of the news from Russia. On the other hand I could not understand why the U.S. project was to be delayed. It was not my business; Hagen knew all the facts and I could only presume that Vanguard was much more nearly ready than Hagen had publicly indicated. Unfortunately, as was soon to be revealed, Vanguard was not only late but nearly a total failure. The U.S. had failed to give the project the priority and support which was necessary. As already mentioned in Chapter 27 I returned to Jodrell Bank in mid-September to be submerged again in the troubles of the telescope, and the question of the earth satellite vanished from my mind.

On 4 October I had no intention whatever of using the telescope to observe the Sputnik. The telescope was full of troubles, contractors were in possession, we could not drive it from the control room, there were few research cables on it, and the radar apparatus which we eventually intended to use on the American Vanguard was not on the telescope neither had we any immediate plans for installing it. We could, of course, easily have used the telescope to receive the beacon transmitter from the Sputnik—the 'bleep-bleep'—but this seemed pointless because these signals could easily be picked up by a cheap commercial receiver connected to a small dipole aerial.

However, my restricted scientific judgement on this failed to take account of many other matters which were soon to descend on me. The telephone bells rang continually. For reasons which I did not understand, the news media in Great Britain seemed to think that the telescope was associated with the Sputnik—or ought to be. The reaction of the people as a whole was extraordinary: a mixture of excitement, admiration, and in some cases fear at this Soviet accomplishment in advance of the Americans.

The *Manchester Guardian* in a long leading article on 7 October expressed the situation admirably.

> Thinkers, technicians, and manufacturers must have been given their fullest scope. Their achievement is immense. It demands a psychological adjustment on our part towards Soviet society, Soviet military capabilities, and—perhaps most of all—to the relationship of the world with what is beyond. . . . What of Russia's war capacity on earth? There little doubt can remain. Mr. Khrushchev was speaking no less than the truth. The Russians can now build ballistic missiles capable of hitting any chosen target anywhere in the world.

On the morning of 5 October (a Saturday) the avalanche of interest was merely the beginning, but I was surprised to find a number of reporters at Jodrell Bank. J. G. Davies was just back from Washington, where he had been attending the Satellite Conference. I hoped that he would bring with him some real information, but although the Russians were at the conference he heard the news only when he landed at Ringway.[1] We discussed the situation and for the reasons given above decided to do nothing with the telescope. He had arranged to attend a meteor conference in East Germany and I told him that I saw no reason why he should not leave as arranged on Sunday. This was just another in the series of appalling misjudgements which I made in those twenty-four hours. Davies was the only one amongst us who was gifted with the ability to think instantly and intuitively about orbits. He was one of the few people in the world at that moment who, with a vast experience of meteor orbits, could have translated this experience into a judgement of the probable behaviour of the Sputnik and its carrier rocket. Today when the positions in the sky of satellites are given with the weather reports in the newspapers it is hard to remember that in October 1957 nobody could predict with any reliability the probable behaviour of orbiting objects. So Davies flew away beyond recall or possibility of communication for the next week and although I did not realize it on that Saturday morning I was to regret most deeply that I had not asked him to remain at Jodrell.

Throughout the Saturday and Sunday a state of siege of newspaper and broadcasting personnel began to develop around my home and Jodrell. I am still uncertain as to why this happened. There had never been any public indication that the telescope had any relation with earth satellites; maybe our Press relations were so bad and our public image in such disgrace that there was a feeling that some vague demonstration of the usefulness of the telescope should be forced. There was also the point that no one else in the country apart from our colleagues with Martin Ryle in Cambridge either seemed capable or willing to give any scientific information about the satellite, its probable behaviour, and the general implication of the Russian achievement.

I suppose that honour and the Press would have been satisfied if we had connected a receiver to the telescope and steered it to receive the bleep-bleep from the satellite, but since the bleep-bleep was coming anyhow from simple receivers on the ground at Jodrell I refused to engage in this kind of pointless demonstration for effect. For me it was

[1] Subsequently it transpired that the news of the launching had been given by an *American* at a cocktail party in the Russian Embassy, Washington, one hour after Davies had left to catch the plane. Amongst the Russians present and attending the conference was Academician Blagonravov, the head of the Soviet rocket establishment—a person whom it was presumed was closely associated with the Sputnik.

either the radar experiment already planned for the telescope or nothing, and with our apparatus and the telescope in the condition of early October any rational judgement would have placed the possibility of doing this experiment many months away.

On Monday morning the situation changed dramatically. There were two major reasons: one concerning my own outlook and the other involving Husband. As far as I was concerned a number of telephone calls from various places in London informed me that there was no defence or other radar in the country capable of detecting the carrier rocket. Although I had spent so much of my career on military radar I had been out of touch during the last few years of telescope crisis and I simply could not believe that those responsible had so neglected the situation. Indeed, my inactivity during those few days had been founded on an assumption that many radars in the country would be detecting the carrier rocket and that any use of the telescope for a demonstration of this possibility would have been regarded with disfavour and pity by my colleagues in charge of the defence radars. Before the end of that day I had learnt with incredulity that, at least in the free world, not a single radar had succeeded in locating the carrier rocket—and this was the rocket of a Russian I.C.B.M.! The miracle was waiting to be grasped. With the vast gain of the telescope even the low power radar transmitter which we were using on the moon experiments should be able easily to obtain a radar echo from the carrier rocket.

The installation of this lunar equipment in a few days would have been a formidable task even if the telescope was fully operational and under our control. Husband had informed a member of the Press over the weekend that the telescope was 'operational', and he had expressed considerable annoyance to me that as far as he and the contractors were concerned we had not demonstrated its potential by receiving the satellite signals with it. When the situation changed on the Monday I told him that my attitude remained the same over the bleep-bleep, but that we would attempt the radar experiment if he, for his part, would summon Brush, Dunford & Elliott, and the other parties necessary to complete the work so that the instrument could be driven through the computer from the control room.

Work which had previously been thought to take months was then completed in forty-eight hours. The Brush engineer who had refused to continue with 'University people breathing down his neck', now returned and completed the job with University people and the whole pack of publicity people breathing down his neck. The change in attitude under this stimulus was fantastic. At 6 p.m. on Wednesday 9 October the servo loop to the telescope drive was closed and for the first time the instrument was moving automatically under remote control from the control room.

Meanwhile we had to face enormous scientific problems consequent upon the decision to attempt the radar experiment. The entire equipment weighing many tons was in a laboratory hundreds of yards from the telescope. To move it bodily on to the telescope at short notice was out of the question. Our only hope was to connect it by transmission lines across the intervening field and then by flexible cable on the telescope itself. The cable was in our schedule for telescope equipment but we had never been able to afford it and we certainly had no money available to purchase it at that moment. By this time the Press were around in force and maintained a minute scrutiny of our activities. The trouble over the cable did not escape them and the resulting publicity had the fortunate circumstance that Sir John Dean of the Telegraph Construction and Maintenance Co. Ltd. most generously offered to send us any high-frequency cable we required instantly. The cable arrived at Crewe Station on the Wednesday evening as the servo loop on the telescope was being closed.

Our first ambition was to test the telescope with this lunar transmitter connected to it, to see if we could obtain radar echoes from the moon. By the Wednesday evening the telescope was moving automatically on the correct coordinates, the transmission lines were across the field, and on the fixed parts of the telescope structure, the aerial had been mounted on the aerial mast of the telescope, and we merely wanted this flexible h.f. cable to join the various sections of the transmission lines. Dr. Stanley Evans offered to collect the cable from Crewe. Ninety minutes later he returned without the cable, having had great difficulty in finding where it was at Crewe and then realizing that a few hundred yards of heavy cable on a large drum could not be moved single handed or transported in a small car. We dispatched him a second time with a suitable van and helpers, and late at night the necessary joins were made in the transmission line. A few minutes after midnight the telescope was transmitting to the moon and strong echoes were received on the cathode ray tube.

The entire place was seething with reporters, live TV was being transmitted, and the noise and glare of light and publicity made any scientific task ten times more difficult. At that stage the B.B.C. alone had many more engineers on the site than I had on my whole staff at Jodrell Bank. They captured the excitement and the first radar echoes obtained with the telescope.

Echoes from the moon satisfied us that the equipment and the telescope were performing satisfactorily, but we had far to go before we could hope to achieve a radar contact from a fast-moving carrier rocket whose position in the sky was only vaguely known to us. On top of all of these problems I faced a severe split amongst my own staff. It happened to astronomers all over the world. Some were enthusiastic about the opening of this new age, others saw it as an interference to their work. The former immediately saw new possibilities for scientific research, the latter complained that the vast sums of money involved would be better spent on more telescopes and equipment. So it was in the local Jodrell community. I do not for one moment blame those who retired to their homes or offices disgruntled that we had temporarily taken the telescope away from the slow and painful processes of initiating its researches in the conventional way, but I am profoundly grateful to those half dozen who stood by me in those historic days. Amongst these were Dr. Stanley Evans[1] whom I have just mentioned in connection with the cable. He had been to Halley Bay in Antarctica as a member of our I.G.Y. expedition and with his younger namesake, but no relation, Dr. J. V. Evans[2] formed the team using the lunar radar. They remained imperturbable and highly efficient under this extraordinary pressure. It was in their small laboratory, almost monopolized by the cathode ray tube display, that we saw the first lunar echoes during the Wednesday night. It was with this lunar apparatus that we planned to locate the rocket, but our success with it was short lived. Almost immediately some obscure fault developed which—probably because of the extreme weariness of all concerned—could not be located.

Fortunately, although half my staff had disappeared into oblivion, the few who remained were full of enthusiasm. Amongst these was Dr. J. S. Greenhow[3] whose research programme involved simultaneous radar and photographic observations of meteors from a site on the cliffs near Abersoch (N. Wales) and Jodrell Bank. On Tuesday 8 October when Greenhow realized that we were proposing to connect the lunar radar to the telescope he came to me holding an envelope on the back of which he proceeded to demonstrate that if we could place his small meteor transmitter in the swinging laboratory of the telescope

[1] Now Director of the Scott Polar Research Institute, Cambridge.

[2] Now at the Lincoln Laboratories, U.S.A.

[3] Dr. Greenhow later went to the Royal Radar Establishment, Malvern, but to everyone's sorrow he died at an early age a few years later.

Three young research students working with J. S. Greenhow spared no effort to make his proposal work. They were J. Davis (now in Australia), J. Hall (now at the Radio and Space Research Station, Slough), and E. L. Neufeld (now with I.B.M. (United Kingdom) Ltd.).

Until the return of J. G. Davies from Germany, J. Davis offered to help me work out the probable orbit of the rocket and direction of best look for the telescope. We did this in my office with a globe, string, and a map of the world.

then we stood a better chance of locating the rocket than with the lunar radar. His arguments were perfectly sound. Although his transmitter was less powerful than the lunar radar one he could place it in the swinging laboratory, immediately under the telescope bowl, and thus avoid the severe losses in the long cables and transmission lines of the lunar equipment. There was another important point. His apparatus worked on a frequency of 36 Mc/s compared with the 120 Mc/s of the lunar radar and hence the beam of the telescope would be nearly four times broader, giving us a far greater chance of success in the search. Of course I agreed immediately, and by the morning of 10 October we had this apparatus on the telescope in addition to the lunar radar. With the latter equipment still out of action we made our first search for the rocket during the night of 10 October with this meteor equipment. We found nothing, and later discovered that the receiver had gone out of adjustment during the transfer.

When the rocket was above our horizon on 11 October we tried again. This time the equipment was clearly functioning. The cathode ray tube was full of the radar echoes from meteor trails and although there was absolutely nothing to guide us as to what an echo from a rocket would look like we were reasonably satisfied that one of the responses was at such a range and of such a character that it was a response from the rocket. The next evening (Saturday 12 October) there was no doubt at all. Just before midnight there was suddenly an unforgettable sight on the cathode ray tube as a large fluctuating echo, moving in range, revealed to us what no man had yet seen—the radar track of the launching rocket of an earth satellite, entering our telescope beam as it swept across England a hundred miles high over the Lake District, moving out over the North Sea at a speed of 5 miles per second. We were transfixed with excitement. A reporter who claimed to have had a view of the inside of the laboratory where we were, wrote that I had leapt into the air with joy. I cannot remember whether this is correct, but I do remember turning on my elder son who had begged to be present in a corner of the laboratory, and who was at that moment showing more interest in the arts than science at school, with the words: 'If the sight of that doesn't turn you into a scientist, nothing ever will.'[1]

By this time it seemed to me that the eyes and ears of the whole world must be on Jodrell Bank. We had no press officer, no public relations officer or any means of dealing with this situation. However, within these few days the entire atmosphere changed, and the friendliness of all of those present enabled us to survive and succeed in this complex

[1] Whether as a result of this event, or other influences, I do not know, he subsequently graduated in geology.

public and technical situation. Nevertheless it was impossible for me to answer questions endlessly day and night and at the suggestion of the Press Association I agreed to give two Press conferences every night at 6 p.m. and 11 p.m. Night after night both before and after our first echoes our lecture room was packed with reporters and cameramen. Our families must have produced several thousand cups of tea throughout those nights: no reporter dare leave the vicinity and the problem of sustenance for ourselves and the Press became difficult. Our small canteen, although accustomed to leaving meals out to be eaten in the middle of the night, was overwhelmed. One night I found the following note from the lady who was then in charge of the canteen: 'Dr. Lovell. I am very sorry to say there were no Eggs left in the canteen when I came in this morning. I left eight meat Pies for Supper. I am also sorry to say I have no Bacon left as we only have a 1 lb a week and it is due tomorrow.' But friends turned up unexpectedly with bottles and the local W.V.S. offered their services. The whole community had been stirred into a frenzy of excitement and appreciation. J. G. Davies returned from East Germany on Saturday 12 October having had no inkling of what was happening at Jodrell. He recalls that he was staggered at the change and found difficulty in getting in because of the collection of T.V. and broadcasting vans.

After two weeks of this life the pressure slackened, but only temporarily. The launching of Sputnik II on 2 November with the dog Laika on board stimulated the interest to even greater heights. The Evans' pair were distraught with annoyance at the failure of their lunar equipment, but after a week they found and repaired their troubles and had their reward with the successful location of the carrier rocket at great range. At their first attempt a few minutes after midnight on 16 October they succeeded in producing a radar echo from the rocket when it was over the Arctic Circle at a distance of 1,500 kilometres. It was this equipment which observed the final orbits of the carrier rocket as it burnt up in the atmosphere in the early hours of 1 December.[1]

On 29 October in the House of Commons the Prime Minister (the Rt. Hon. Harold Macmillan) said, 'Hon. Members will have seen that within the last few days our great radio telescope at Jodrell Bank has successfully tracked the Sputnik's carrier rocket'.[2] By any rational human standards our troubles should have been at an end.

[1] The critics on my staff who were against the use of the telescope for these purposes maintained that we were using it for publicity and not for any programme which had a scientific justification. They were wrong. Important scientific results were obtained which were published immediately in *Nature*, vol. 180, p. 941, 1957, and in Proc. of the Royal Soc. A., vol. 248, p. 24, 1958. The observations on Sputnik II were published in Proc. of the Royal Soc. A., vol. 250, p. 367, 1959. [2] *Hansard*, vol. 525, No. 156, p. 31.

31

The search for a financial solution

If our troubles had been merely financial I think that a solution would have been close at hand in that fantastic autumn. Jodrell Bank and the telescope was scarcely out of the news for months. We began to gather public sympathy and enthusiasm as it became a symbol of at least some British contribution to the 'space age'. Placards appeared, issued by the Government, featuring the telescope as 'Britain's great achievement'. Letters descended on M.P.s in their hundreds and there was lobbying at the highest levels.

Unfortunately, issues of deep principle had become involved, as well as money. False statements damaging to Husband were in the public record; in turn he had threatened me and the University with legal action for damages. The Treasury had presented the University with impossible conditions for making even an additional grant for the telescope (see Chapter 26) and the promises extracted by the Public Accounts Committee from the D.S.I.R. and the Treasury made it impossible for the Government to settle the bill and close the matter honourably, however much it may have been politically desirable to do so.

Two years previously I had made a too impetuous start in attempting to raise money to clear the debt (see Chapter 23). Now I could not even have this release of activity to help myself and others. I was torn asunder: at one moment elated by the success of the telescope, the next, sunk in the morass of legalities with my whole career at stake.

Further, my inexperience in dealing with the Press and the lack of anyone to guide or advise me in those weeks of extreme pressure led me to make remarks which would have been better left unsaid—justified though they may have been. For example, at the packed Press conference on Sunday night 13 October 1957 I was able to show on the screen the radar echo from the rocket as we tracked it over England the previous evening. It was a marvellous photograph. We were still highly excited and so were the Press. I still have a cutting from the *Manchester Guardian*'s front page of the following morning (14 October). It is full of praise for the telescope, for us, and for the photograph. Alas, at the bottom of the page: 'At the end of last evening's press conference Professor Lovell said the Government had been treating him and his colleagues like criminals because there had been overexpenditure on Jodrell Bank . . . perhaps now the Government would understand its importance.' Two weeks elapsed and then under an imposing London letterhead: 'Dear Lovell, My attention has been drawn . . . we have gone to great lengths to try and help them [the University] in their financial difficulties, but statements of this kind issued in the public press naturally do not make it any easier to conduct these negotiations in a reasonable way. I have personally spent etc. etc. . . .' Richly deserved, yes, but what a difference even one word of appreciation from London on the achievement of the radar track would have made. Throughout all those months there was none, just suspense and criticism. The vision was submerged in the tangles of bureaucracy.

The dichotomy was apparent in the newspapers. For example, after the eulogies of early October the *Manchester Guardian* of 23 October featured JODRELL BANK COSTS followed by columns on the detailed minutes of the evidence given to the P.A.C. on which the August report was based,[1] and in the same paper there was another item:

TWO GIRDERS OF TELESCOPE BENT. IS IT OPERATIONAL?

Two steel girders essential to the stability of the Jodrell Bank telescope have been bent and are now having to be replaced by stronger ones . . . they were bent when the bowl of the telescope was tilted at a considerable angle to the horizon during calibration trials some weeks ago. . . . their present twisted appearance is visible some distance from the telescope. In the circumstances it is inevitable that there should be some confusion as to whether the telescope is now fully operational.

This shows how ready the Press was to pounce on us even after the successes of a few days previously. The bent girders were minor members in one small framework. But only five days later there was another side to the picture:

TELESCOPE TRIUMPHS

. . . the Jodrell Bank telescope handsomely justified its existence—deformed steel members notwithstanding—by detecting the Soviet rocket at a range of more than 1000 miles. This performance has probably never been equalled by a radar instrument.[2]

Until the end of the year the work of completing the telescope dragged on. On 2 October I had made an entry in the diary, then in the furore of the Sputniks I abandoned it entirely. On 1 January 1958 I made the

[1] The minutes of the evidence were not published until 22 October 1957.

[2] *Manchester Guardian*, 28 October 1957.

last entry of all to summarize in one paragraph the happenings of the previous chapter. Then I wrote:

> All this time the steel workers were repairing the bent girders and it was with the utmost difficulty that we managed to get limited use of the telescope. The weeks dragged on and still the telescope remained unfinished. Now the only outstanding job is the trueing up of the Brush mechanisms. . . . Goodall is leaving today and it does at last seem that only days separate us from a reasonably full use of the telescope—that is apart from the severe organizational problem of the hand-over.

The year ended as it had been throughout—in a ferment. The problems of hand-over were not only those to be expected, but were exacerbated by the natural desire of Husband to reserve his position until the financial and other issues had been dealt with. It must be admitted that without experience to guide, we had not made proper provision for the necessary engineering staff to be responsible for the telescope. Husband was asked to give advice and produced a document with which I could not entirely agree. At the time I was annoyed that certain parts of the document seemed to be laying down rules about matters which I thought could only be the internal concern of the University. But it is only fair to say that now, on re-reading this document, I find it hard to understand why we did not accept it as commonsense. Today our organization for dealing with the engineering aspects of the telescopes is close to that proposed by Husband in the autumn of 1957. Fortunately I was gently but firmly restrained by the Vice-Chancellor from the various resignations which I threatened. The University sent the document to D.S.I.R. because it proposed the employment of many more men and a senior engineer, for which money would be needed. Towards the end of the year I wrote: 'the D.S.I.R. were believed to be trying to get an assessor from the National Physical Laboratory to sort out the difficulties. In the meantime Renold has intimated that I might not be allowed to use the telescope.' Of course this message coming over the phone nearly drove me crazy. It does indicate the inconceivable tangle of our affairs that at that moment, after nearly ten years of struggle, with the Sputnik epic recently accomplished and with the full use of the telescope within sight, I was being threatened with excommunication from my own laboratory.

The rest of the world took a different view of our affairs. In *The Times* of 1 January 1958 there was a heading: 'THANKS FROM MOSCOW FOR PROF. LOVELL . . . has received a telegram from Moscow with New Year wishes. The telegram adds: "Every success in your work. Best thanks for satellite operations'."

The months of tension

The drama surrounding the association of the telescope with the Sputnik had completely changed the general attitude towards it. These events had not in themselves, however, created the conditions under which a settlement was possible. Before any progress whatsoever could be made there were two major issues which had to be settled. Firstly some formula had to be found by which the Treasury's conditions respecting the award of the additional grant to the University by D.S.I.R. could be met. Secondly the wrong done to Husband and his firm by the erroneous statements in the P.A.C. report had to be redressed. The two were interconnected and tangled. The University could have met the Treasury's conditions by moving legally against Husband. It was clear that a single step in this direction would have been instantly countered by Husband suing me for damages for allowing the P.A.C. published record to stand.

There were many influential people who did not know of these detailed entanglements and who began to make approaches at high level to persuade the Government to settle the debt. These movements still further increased the vast difficulties facing the University and D.S.I.R. in the search for a formula. The situation needed and was treated with the greatest diplomacy. Again and again the Vice-Chancellor urged on me the utmost caution. When we were literally submerged by the Press because of Sputnik, the P.A.C. minutes were published. The Vice-Chancellor wrote to me on 23 October 1957:

The Press as you know is full of the evidence given before the Public Accounts Committee and our negotiations with the D.S.I.R. and the Treasury have reached a highly critical stage. Because of this I want to enjoin you to be extremely cautious in any statement that you make to the Press. Say as little as you possibly can, short of losing their good will.

In January when I sought the Vice-Chancellor's advice about pressure which some of my friends were bringing at the highest levels he wrote to me on 30 January 1958:

I am hoping we are, within the next fortnight or so, going to solve major problems concerning the telescope. It is true that we have been hoping for this for months but I think there are one or two propitious signs. In view of this would you kindly hold your hand from contact with — or anyone else about unorthodox approaches to the financial powers. I think we must work at the moment through D.S.I.R. and no one else.

On 2 February Sir Raymond Streat wrote to me:

Just a line to assure you that I am fully aware of the urgent importance of the finance side of the telescope and trying by every means in my power and every day that passes, to get ourselves free to act. I sent an acknowledgement of your letter [about the other direct approaches to the Government] of a week ago through Miss Miller [the Vice-Chancellor's secretary] and raised it at the Officers' meeting. But I do want you to feel quite sure of my unremitting efforts at my end. I feel we should get to a better position within a week or two. Anytime it would help you to talk with me or even just comfort you, telephone and I will drive over.

The crisis was at its height. Two days later the Vice-Chancellor advised me that I might be requested by solicitors to write a letter setting out the fact that I had been in continuous consultation with Husband. He asked me not to do this without consultation with him.

On 7 February I wrote to Raymond Streat to ask if he could see me. On the following Sunday afternoon he came with Lady Streat to see us at my home. Raymond and I had tea together in my study where in the kindest possible way he said that it was his unpleasant duty as Chairman of the University Council to warn me that I must consider myself to be in dire peril. I asked him if he could be more explicit and he replied that it was the opinion of the University that a writ might be served on me in the near future for a third of a million pounds. I asked, in view of the fact that I did not possess such a sum of money, what would happen. He answered that he was sure the University would bear the costs of the legal action but as long as the P.A.C. record was allowed to stand they were afraid that the case would be lost. In this case failure to pay would mean imprisonment.

The memory of this scene, of Raymond Streat's sympathy and kindness, of the bleakness of that afternoon are indelibly engraved on my memory. The conflicts and contrasts were beyond belief. In the public mind the instrument had become the symbol of Britain's contribution to space, the contrast heightened by the arguments centred around Blue Streak and the truly vast sums of money at issue there. Apart from its work on the Sputniks the telescope was then beginning its astronomical researches. Its performance was without parallel. It was the only major scientific research instrument which the U.K. had produced since the war. Thousands of hoardings and advertisements carried its image. I showed Raymond a copy of a message which I had received from the Prime Minister's office the previous day. It assured one of my contacts who had been pressing for a settlement in high circles, that a settlement 'under certain conditions' had been agreed. There was still no acknowledgement that the 'certain conditions' made any settlement impossible. I do not think that many people either then or subsequently realized that this conversation in a Cheshire village on that February afternoon revealed that the entire enterprise was in the greatest of all its perils—and this at the very moment of its achievement.

This final catastrophe did not happen because the inexhaustible patience, skill, and commonsense of those involved slowly but surely found areas of agreement through which progress could be made. The vital step was the withdrawal of the previous evidence before the Public Accounts Committee.

In the week following Raymond Streat's conversation with me, Husband submitted a memorandum to Sir Harry Melville, summarizing the factual evidence of our close association and concluding:

> Messrs. Husband and Co. hope that the Committee of Public Accounts will think that, in the light of the representations which have been set out herein, the statements contained in paragraphs 62 and 64 of their Report, and which have been so damaging to Messrs. Husband and Co., do not represent a true view of the actual sequence of events at Jodrell Bank nor of the conduct of the Consulting Engineers.

On 28 February Melville wrote to the Clerk of the P.A.C. conveying this memorandum.

The P.A.C. asked the Vice-Chancellor and Rainford to give evidence before them on 18 March. By coincidence I was addressing the Parliamentary and Scientific Committee in another room of the House at the same time. We emerged simultaneously. They were unhappy and not hopeful as to what the Committee would say in its report.

The situation remained tense and our anxieties were exacerbated by the demands of the United Steel Structural Co. for a settlement of £124,405 in excess of the latest estimates on which our deficit of £260,000 had been assessed. These additional claims were vigorously contested by Husband, who advised us in April that the United Steel Structural Co. were considering abandoning their claims by way of giving a substantial contribution to the radio telescope. The Bursar was concerned that this solution to the difficulty would weaken our case with the Treasury and D.S.I.R. in respect of the 50 per cent. contribution to the agreed deficit of £260,000. It was within the framework of this tense atmosphere, uncertainty over the nature of the P.A.C.'s report after the 18 March meeting, with the probability that the University still might not get a contribution without satisfying the agreed Treasury conditions, and with the certain knowledge that if everything turned out in the best possible way, we still had to find £130,000, that the Vice-Chancellor summoned the officers and all concerned to discuss the problem at dinner in the Staff House on 30 April 1958. The members of the Site Committee (Renold, L. P. Scott, Matheson), the Chairman of Council (Sir Raymond Streat), the

Treasurer (Mr. A. V. Symons)[1] and his deputy (Mr. R. B. Barclay), the Vice-Chancellor, the Bursar, and myself were present.

For the dinner Raymond Streat had tabled an 'Outline of Telescope Policy', a masterly summary of the complexities which then faced us and for which we had been summoned on that Wednesday evening to find a solution. His memorandum began:

> The total problem of policy has the following facets:
> 1. The deficit on Capital Account: when and how to reimburse the University
> 2. Revenues for running the telescope (a) operation and supervision of mechanical aspects, (b) scientific programmes
> 3. Relations with Mr. Husband
>   (a) financial (b) legal (c) prestige considerations.
> 4. Considerations affecting the welfare of the University internally, including: (a) consequences of capital deficit on University programmes generally (b) consequences on morale and contentment of other departments
> 5. Considerations affecting the University's external relations with (a) D.S.I.R. (b) U.G.C. (c) Public opinion (d) Opinion in other universities (e) Treasury (f) P.A.C.
>
> Whilst the chief practical issues (i.e. Capital Deficit, arrangements with Mr. Husband, and running the telescope, 1, 2 and 3) are of great and pressing importance they are extensively bedevilled by the intangible issues. The ideal policy, or plan of action, would be one which held out hope first of clarifying every facet of the total problem and then of proposing steps likely to cope with each and all of them. It seems essential that the University should know what its own version of the whole affair really is and what its aims are now. It may prove essential to be able to inform our own colleagues and those, externally, with whom we are involved in this matter as to our views and intentions. It may be that only a publicly available document will meet this need since a secret document will give rise to many suspicions and much controversy.

Raymond Streat's document sets out the seemingly inextricable entanglements with which the affair of the telescope had become involved. For me, as the person who had been the cause of the trouble, and who now had to justify the scientific importance of the telescope, 2, 'Revenues for running the telescope', had become a pressing issue. The University was unable to increase the small grant which I had for research before the days of the telescope, the normal avenue of approach to D.S.I.R. for special funds in such circumstances 'for researches of timeliness and promise' was barred. Equipment which we urgently

[1] He succeeded Sir Raymond Streat as Chairman of Council of the University in 1965.

needed for minimal researches with the telescope could not be purchased. Furthermore I was enormously burdened by the sudden descent of the responsibilities for 2(a). The engineers had left the site, the resident engineer had departed. We had an enthusiastic and willing half dozen engineers of our own, we were trying to obtain people whom we could train as operators of the telescope for the control room, but this staff was hopelessly inadequate for the job at hand and we had no senior engineer who could take the immediate responsibility for these tasks. Apart from our huge debt it was therefore a question of urgency that our affairs should be settled so that we could make progress with the negotiation of funds for the operation of the telescope in all its aspects.

Apart from Raymond Streat's document, the dinner party also had papers before it from Sir Charles Renold giving some comments on the situation and a memorandum from Rainford on the Structural Steelwork Costs in which he showed how the United Steel Structural Co.'s current claim for £417,016 compared with the agreed October 1955 estimate of £292,611. In this memorandum he stated that

> as a result of prolonged meetings with the United Steel Structural Company, he (the Consultant) had agreed with them that the weight of steel supplied and erected was 1750 tons. This was the weight of steel allowed for in the 1955 estimates. Whilst the weight of steel is therefore not in dispute, the United Steel Structural Company contest that the change of design caused considerable erection difficulties and, in consequence, additional expenditure. This is the main issue in dispute between the Consultant and the United Steel Structural Company.

As far as I am aware there are no minutes or reports of the discussions at that dinner; I have none myself, and depend on memory alone for the recollections of an evening which was decisive to our fortunes. In the prevailing situation it was hardly to be expected that there would be unanimity or that any decisive line of action would emerge. From some there were strong feelings that the only solution for the University was to accept the Treasury's conditions and secure the D.S.I.R. grant, and then somehow to raise the balance whatever the consequences to the individuals involved. I knew well enough that as far as my future with the telescope was concerned, and probably the future of the telescope itself, this would have been the end of the road. After many hours, when papers were being folded up, and this view was in the ascendancy, the Treasurer, A. V. Symons, quietly said 'Vice-Chancellor, I suggest we take no action, time is the great healer'. At least I remember that closing scene of the occasion when the entire future of Jodrell Bank was at risk.

As the days lengthened so we slowly moved away from the lip of the precipice. On 24 July 1958 the Public Accounts Committee presented its report: '. . . In view of this further evidence, it is clear that the evidence given to the Committee of last session was gravely inaccurate and misleading, and that there was in fact the fullest collaboration on scientific and technical matters between the consultants and the University Professor.'[1]

The greatest single impediment to action was removed. The University was at last released from its shackles and moved into action. With the receipt of the additional £130,000 grant from D.S.I.R. the immediate target was to clear the remaining capital deficit of a similar amount. There was, of course, the additional anxiety of the United Steel Structural Co.'s claim which, if accepted, would have required us to raise an additional £124,405. Fortunately, this great additional burden was avoided by the generosity of the United Steel Companies Limited. Acting on Husband's repeated advice the University arranged to extend an invitation to Sir Walter Benton-Jones, the Chairman of the United Steel Companies. He came to Jodrell Bank on 11 July 1958 and the favourable situation which developed from this visit was later acknowledged in a fitting manner by the Chairman of the University Council (see p. 207).

By the autumn of 1958 the entire atmosphere had changed. The telescope was again involved in international affairs of great moment (see Chapter 32). At last I was free to seek financial aid from D.S.I.R., and on 21 October 1958 Greenall of D.S.I.R. wrote to me: 'This informal note is just to let you know as quickly as possible that we have now had some welcome news from the Treasury about your recent application for an equipment grant for the radio telescope.' On 24 October I wrote to Rainford about 'my visit to Cawley at D.S.I.R. yesterday. He wishes us to apply again for the running and maintenance costs of the telescope . . . he thinks that this can now be got through . . . I was surprised to find such a marked change in the atmosphere and I think that we may receive substantial assistance on the research side from another scheme which is now being developed.'

Quite apart from the entanglements of the capital deficit my own new and emerging anxieties in the running of the telescope were thereby eased, and during the summer the appointment of Commander R. F. Tolson,[2] as Chief Engineer removed from me a heavy burden. The

[1] Third Report from the Committee of Public Accounts Session 1957–58, para. 60, p. 20.

[2] Commander R. F. Tolson, D.S.C., A.M.I.Mech.E., R.N., recently retired after thirty years service in the Royal Navy of which a large proportion had been spent in the submarine branch.

stage was set for the final appeal. Its preamble was the superb speech to Court by Raymond Streat on 12 November 1958. In the following passages he simultaneously summarized our anguish of the past years and epitomized our achievements and hope.

> Before concluding I wish to make special reference to one topic not mentioned in the text I am presenting. That is the topic of the Radio Telescope at Jodrell Bank which has been mentioned, I imagine, in every newspaper in the world, or very nearly so, in the last few days, because of the American lunar probe experiments. You might think it strange if a subject of such immense interest to the world at large at this present moment were not to be mentioned at a Court Meeting of the University to whose credit this unique achievement stands.
>
> I acknowledge that we have been subjected to criticism as regards the financial and administrative arrangements at various stages of the history of the enterprise. We at the University have resisted every temptation to enter into public discussion of such criticism which might, in any way, diminish, or lead to misunderstanding concerning the immense achievement at Jodrell Bank. For we believe that the Radio Telescope cannot be described in lesser terms. It is an immense achievement whether viewed as a piece of scientific equipment conceived by Professor Lovell or as an unparalleled feat of engineering designed by Mr. Husband. Every day that passes emphasizes these facts and the time is at hand when all due acknowledgments to the help we have received from many sources, may be made.
>
> In the meantime there are one or two things I wish to emphasize today. We have received continuous encouragement and assistance from the Press. Its support has been, at every stage, invaluable. Unfortunately, however, there has been some suggestion that British industry has been unhelpful. The facts are far otherwise. Although we have not yet made any formal appeal for the funds which will enable us to clear off the deficit on the cost of the instrument, we have already received a sum of £10,000 from Henry Simon Limited, and £10,000 from Simon Carves Limited. And throughout the whole undertaking the University has been sustained by the co-operation—almost amounting to partnership—of the many firms engaged on the project and chiefly by the main contractors, the United Steel Companies Limited. That Company's co-operation in the building of the telescope has, I feel sure, caused them on many occasions to rate the final achievement much above purely financial consideration. Indeed we know full well that from the Company's point of view their association with the University in this venture has not been financially profitable but in view of the novel character of the structure and its great importance to scientific developments in this country, they have never failed to meet our requests. We are greatly their debtors on this account and in my opinion the nation and world of science are also in their debt.
>
> These remarks of mine will make clear that the University already has cause to be grateful to British industry in this matter.
>
> As to the criticisms made at various points I think the Court of Governors

will feel as I do that the whole story reflects a brilliant light of credit on all concerned. The criticisms arose from the fact that the Public Accounts Committee, in the perfectly proper execution of their public duty, began to discuss in public some aspects of the finances which the University was not at liberty to deal with in equal publicity because of the complexities of its legal and honourable commitments to other parties. This made it possible for the public to think the finances had in some way been handled inadvisedly. This was not really so. True, there were developments which nobody foresaw at the outset. But I venture to assert that no comparable pioneer enterprise has ever proceeded from conception to fulfilment without at least some unexpected developments. The fact is that Britain is first in the field with a new scientific instrument of fantastic quality, at a total cost far below what others will be soon paying for similar instruments and Britain's contributions to a new field of knowledge will in consequence for ever redound to our credit. It is true we at the University might have taken different decisions as each of the several unforeseen developments in our problems, scientific or financial, emerged. So might the generals in command of any military campaign. But the proof of the pudding, for generals, is the success of the campaign and the proof of the pudding for the University is success in achieving the original conception in time for it to play an irreplaceable part in historic scientific events.

As to the amounts still outstanding in the total cost I am sure we will be able to attract satisfactory contributions. Before the end of this month we shall be addressing an appeal to those whom we believe may be disposed to help us. We feel confident that there will be many who will be extremely keen on this great national achievement, resounding round the world from the Brussels Exhibition to Cape Canaveral and, no doubt, Moscow and will be ready and willing to help the University to wipe off what is relatively a small debit balance and so free us for scientific work of the first order of importance and urgency. Anybody who feels moved to give before being formally asked to do so and believes with the soothsayer that he gives twice who gives quickly will find our Treasurer already primed with a book of receipts.

More will fall to be said about this Manchester University achievement in coming days. Meanwhile we can all rejoice in an accomplishment which will for ever reflect honourably on the vision and enterprise of the University.

On 28 October 1958 Lord Simon came to Jodrell. He came frequently and this may not have been his last visit, but although I saw him at his home on several occasions subsequently, I like to remember that day as his farewell to the telescope.[1] As usual his black pocket-book came out of his pocket and with pencil poised he asked the question so frequently asked in the past. What were my troubles. There were no real ones except the money. I was not to worry, that would sort itself out, 'but technically what about the telescope?'—'Oh that's marvellous'. The pocket-book snapped to without a note 'Good, I do like success'.

[1] Lord Simon of Wythenshawe died on 3 Oct. 1960. An account of his life and many activities is given in the book by Mary Stocks *Ernest Simon of Manchester*, Manchester University Press 1963.

32

Project Able—the American moon rocket

THE deep gloom of the early months of 1958, occasioned by the unfolding of the events described in Chapter 31, was touched by two events which sustained my hope for the future. Less than two weeks after Raymond Streat's visit to me in February when all life seemed to be a wreckage I received a letter from General Sir Ian Jacob (then Director-General of the B.B.C.) inviting me to give the Reith Lectures that autumn. I do not know by what processes of consultation the B.B.C. choose their Reith lecturers. To those who reached a decision to invite me I remain profoundly grateful for this demonstrative reaffirmation of faith in me as an individual and in the project which was nearly destroying me. The lectures were born from the conflict of the agony of the telescope and the global and universal events for which it was created.[1]

The second event, occurring just before Easter and a few weeks before the dinner party of the previous chapter, was destined to have a major influence on our future. I received a secret and highly confidential message that Colonel L of the United States Air Force desired to visit me. The arrangements for this visit were so confidential that I can find no record of it in any of our files or diaries. I recall that I was asked to meet him personally off a certain train at the local Goostrey railway station. I did this without the slightest indication of the nature of his visit. We walked off the station remarking on the beauty of the spring morning as though we were out for a day's walk and continued in this frame until we reached my office. He then requested me to shut all windows and lock the doors, and the real conversation then began in a scarcely audible near-whisper. I was astonished at the message he brought and by his request. It must be remembered that whereas Sputnik was launched on 4 October 1957 the Americans had not succeeded in launching their first earth satellite until 31 January 1958.

[1] *The Individual and the Universe*, the Reith Lectures Nov./Dec. 1958, subsequently published by Oxford University Press and Harper & Row 1959.

The carrier rocket was an Army missile. Col. L now informed me that the Air Force had decided that they could use their Atlas Intercontinental Ballistic Missile to launch a rocket to the moon; further that they could be ready by August but had no means of tracking it. Would we use the telescope for this purpose?

I was amazed at this indication of progress following only a few months after the somewhat supercilious and unrealistic attitude to the Sputnik. It remains a telling indicator of the immense vigour of the American reaction to Sputnik and also of the intense rivalry between the armed services at that juncture before the creation of the National Aeronautics and Space Administration.

The request, which was the reason for Col. L's transatlantic flight, suddenly appeared before me like the vision of another miracle which, if achieved, would surely mark the end of our troublous times. Although as the director of a University research project I was free to decide how to use the instruments at my disposal, this was clearly a case where international relations of an unusual character were involved, and particularly in view of the troubles surrounding the telescope I told Col. L that I ought to seek some formal authority for engaging in this exercise. He replied that discussion was out of the question, that an instant decision was needed, and that he could not allow me to break to anyone the confidential information which he had conveyed. While he was speaking I saw the solution. I said that I would agree to join in this exercise as part of our cooperative international effort in the I.G.Y. programme. I had funds from the Royal Society for running the telescope for certain items in connection with the I.G.Y. so that if questions were asked the answer would be clear provided that he accepted the condition that the U.S.A.F. made no attempt to pay us for our work. He agreed. From our point of view the arrangement left us free of the tangled obligations and negotiations which at that time would have stifled the project. I knew too that the concept, in the spirit of the I.G.Y., would be regarded as praiseworthy. We then discussed details. Col. L said that a complete trailer-load of tracking equipment would be flown over from the Space Technology Laboratories in Los Angeles, about one month before the launch, together with a small party of technicians. In the meantime I was to make sure that the necessary supplies and means of connecting to the telescope would be available.

Our return to Goostrey station occasioned no comment and within hours Col. L was again in Los Angeles, fully satisfied with his mission. For three months the arrangement made in my office remained a complete secret. Apart from a few members of my staff it would have remained so until the day of launch but for an eagle-eyed reporter who noticed with surprise one day in July an enormous trailer on the road from Burtonwood, with the container marked in huge letters 'Jodrell Bank, U.S. Air Force, Project Able'. Then—

Manchester Guardian July 25 1958

JODRELL BANK JOINS IN JOURNEY TO MOON

The United States Air Force will make a serious attempt to get a rocket to the moon in the next few weeks. . . . The radio telescope at Jodrell Bank will be used as an essential tool in the tracking of the rocket after its launching from Cape Canaveral. This can be inferred from the appearance at Jodrell Bank of a United States Air Force van full of electronic equipment. . . . The van at Jodrell Bank appears to have come directly from the United States and is staffed by three or four American technicians. . . . Its presence there can be explained only if it is proposed to use the Jodrell Bank telescope as a means of following the broad outlines of the path followed by a rocket. . . . The Jodrell Bank telescope is uniquely suited to this task . . . That the United States Air Force should have fallen back on the use of Jodrell Bank for this purpose is a telling tribute to the versatility and power of the great telescope.

From my own personal point of view the proposed launching date of mid August could not have been worse. It had been my intention to participate in a symposium on Radio Astronomy in Paris from 29 July to 6 August and then after a day's intermission to go on to Moscow for the meeting of the International Astronomical Union. After the leakage in the *Guardian* we were besieged by the Press. I took counsel with Walter Hingston who promised that he would handle the Press with his D.S.I.R. department. With this release, and because at that time the possibility of conversations in Moscow were of high importance, I decided after many misgivings to carry on with my plans. J. G. Davies was also in Moscow and our part in the operations was placed in the capable hands of J. H. Thomson and J. V. Evans, with the American team responsible for the data collection in their own trailer. Furthermore the proposed launching date had been conveyed to me in strict confidence and any alteration in my travels would have been immediately interpreted by the Press who were following me around like detectives. Indeed, in Paris their attentions became so obstructive that at one stage I was on the point of a decision to return home.

If Hingston had not assumed the responsibility for our external relations on that occasion I do not think it would have been possible for me to leave because a frightful tangle developed over the issue of whether the announcement of a successful launching should be made first from Jodrell Bank. Hingston's memorandum of the developments begins quietly on 6 August: 'Met Col. F (female) at J.B. and she attended meetings with Lovell and Taber (U.S. scientist). F. admitted she knew practically nothing of matter; became clear she was acting as watchdog for U.S.A.F.' On 15 August (two days before the launch) 'Visited Foreign Office. Cable received from Embassy Washington. . . . Foreign Office suggested getting US Embassy to knock heads of U.S.A.F., P.R. organization and mine together. I welcomed suggestion.'

I left for Moscow on 7 August after making a recording with Hingston for him to release if necessary at the appropriate time. The launch date was Sunday 17 August. It was a strange experience to be so tied up with this event and yet so isolated in a foreign capital. Eventually with the help of a friend who was fluent in sixteen languages I spoke to Jodrell Bank on the phone from the lobby of the Moscow hotel.[1] Norman Manners (Hingston's colleague) told me that the probe had been launched but had exploded eighty seconds after launch before it came over the Jodrell Bank horizon. At least I was able to concentrate on the remaining proceedings in Moscow without distraction.

Pioneer I

This first rocket had several reserves which were to be tried as soon as the launch pad could be prepared again. The next was scheduled for October. In the meantime we achieved a more thorough integration of the various pieces of equipment at Jodrell Bank and carried out extensive tests and recordings on an Explorer satellite. The files of the newspapers are eloquent.

The Times October 11, 1958. With the moon in a favourable position again it is expected that the United States Air Force will make their second attempt this weekend to send up a lunar probe . . . Professor Lovell said yesterday that he knew the proposed date for the attempt but he would not disclose it out of courtesy for the Americans. Last night the powerful instrument was to be given an operational test on the orbiting U.S. Vanguard satellite which was launched in March.

[1] The conversation was exceedingly faint and difficult. Ironically the Crewe exchange could be heard with perfect clarity. The trouble was in the final 13 miles of line from Crewe to Jodrell. The efforts to make this phone call occupied the whole of Sunday afternoon and my wife and I were grievously late at an afternoon 'reception'. To our dismay we found it was a full scale dinner (at 4 p.m.) and that our host who was an Armenian proposed our health (expressing relief at our appearance 'like the prodigal son') followed by the demand that we must rise and drink in one act all the healths which had been drunk before our arrival. This amounted to about a half tumbler of vodka!

Evening Chronicle (Manchester) October 11, 1958

OFF TO THE MOON JODRELL TRACKS IT.

America launched a giant rocket—the Pioneer—on a 221,000 mile flight to the moon at breakfast time today. The 88 ft rocket soared off the launching base at Cape Canaveral, Florida, in a flash of brilliant white light within 10 seconds of the scheduled time at 8.42. . . . Prof. Lovell said the giant radio telescope had picked up signals from the rocket at 8.52 a.m. 10 mins after the launching. They were 'most satisfactory'. . . . A tense crowd of reporters and cameramen, including teams from B.B.C. TV and I.T.V. were told everything is going to plan.

Sunday Pictorial October 12 JODRELL SEES THE DAWN OF NEW ERA

Sunday Times October 12 . . . as they sat in the control room beneath the 250 ft high steel mass of the telescope, Prof. Lovell and his assistants seemed to be treating their work almost as a matter of routine. . . . the only sign he showed of tension was that his left hand was covered with red ink.

Amongst the packed crowd of journalists were some distinguished science writers who absorbed the ethos of the occasion as well as the facts. Amongst these were John Davy (the *Observer*) and Michael Frayne (who was then writing for the *Guardian*). Under the date line 'Jodrell Bank 11 October' John Davy wrote in the *Observer* of 12 October:

All day, the Jodrell Bank radio telescope has been following the American moon rocket out into space.

As the wide circular eye of the radio telescope, mounted in its spidery metal frame, moves across the sky, there is a strange sense of direct contact with the Pioneer to enhance this historic occasion.

As the launching time approached this morning nerves became taut. At 7.55 a high shrill note—a curiously appropriate science-fiction sound—was heard from the radio telescope. The wide steel dish began to swing down and round until it was facing straight at the western horizon. (The noise, someone said, was irrelevant—it came from damp brake linings.)

At 8.50 the doors swung open for a moment and we heard a snatch of talk —'up eight minutes now'. At 8.52 there was a stir in the control room, and outside, very slowly at first, the great bowl of the telescope began to move.

The shrill noise filled the air again, as it has now been doing all day at intervals. Faster and faster the bowl swung up from the horizon, turning at first towards the south. The moon rocket was climbing over the Atlantic, passing above North Africa, moving at about 25,000 m.p.h. with all its main rockets already fired.

By 9.30 the telescope was pointing high into the sky, moving very little. The probe was travelling directly away from the earth. The earth's rotation gradually carried its position in the sky back towards the west and down towards the horizon.

Every few minutes, the radio telescope makes small scanning movements, up and down and from side to side. This helps to fix the direction of the radio signals from the rocket to within half a degree.

Signals indicating the rocket's speed, and the micro-meteorites encountered in space are being recorded, and transmitted to the United States. An instrument will measure the moon's magnetic field (if it has any—this, to scientists, is one of the most interesting questions). The transmitter for sending back this information has been designed to work for about seven days. . . . A man-made object is being followed far out into regions of the universe about which a lot remains unknown. For the scientists it is a great moment—and for the romantics too.

In fact it quite soon became evident that the rocket was not in the place where it should have been if it was going to the moon. By the early evening of that Saturday the rocket was over 60,000 miles from earth, but we were locating it more than 10 degrees away from the correct position and it set below our horizon an hour later than we expected. When we located it early on the Sunday morning, it was near the apogee of its orbit and during the course of that day, having travelled just over 79,000 miles into space, it began to fall back to earth. Michael Frayne's description appeared in the *Manchester Guardian* on Monday 13 October, under the date line 'Jodrell Bank, Sunday'.

JODRELL BANK'S TRIUMPH

'We believe the probe is now on its way back to earth' said Professor A. C. B. Lovell, the director of the Anglo-American tracking team here to-night. It had probably reached a maximum height of between 80,000 and 90,000 miles some time to-day.

He expected that it would return to the vicinity of the earth in the next day or two, where it would either plunge into the atmosphere and burn up, or become an earth satellite. Predictions were complicated, however, by the possibility that the probe's retro rocket might be fired.

As Pioneer, hovering in the constellation of Libra, set below the Cheshire horizon to-night, the thin, ethereal screams the radio telescope gives as it moves made a fitting accompaniment to the news of the probe's fate. The feeling here was one of tiredness and strain after Saturday's triumph. But Professor Lovell wanted to make it clear that the Pioneer's failure to reach the moon did not 'take even half an ounce away from the Americans' achievement'.

'If I have a slight look of anxiety,' he said—and he had just that—'it's because of uncertainty as to what's happening, not because of any feeling of disappointment.' The probe would have achieved all its scientific aims except the measurement of the moon's magnetic field, and the televising of the far side of the moon (which he dismissed as 'semi-popular', and which he thought might be replaced by a more valuable televising of the earth). 'People who can launch a rocket like this,' he declared, 'have already got almost everything . . . as a first shot it's wonderful.'

On Saturday afternoon a brilliant rainbow shone in the sky over Jodrell Bank—a medal ribbon for a gallant day in the history of man's achievement. The probe rose above the radio telescope's horizon at 8.52 a.m., just ten minutes after launching, already hundreds of miles up over the Atlantic. The telescope swung to follow it as it reached its zenith over North Africa. Its radio signal was strong and clear; soon Jodrell Bank was receiving it over a greater distance than any radio signal had ever been received before. As the probe went up into space, the turning of the earth swung Jodrell Bank away from it, until it sank towards the western horizon from which it had risen.

JUBILANT

Jodrell Bank was jubilant. Mr John Taber, the head of the American section of the tracking team, admitted afterwards (as reluctantly as if he were betraying a State secret) that when the probe was first located 'I said "We're on!" . . . we were all real happy. I saw an awful lot of smiles on people's faces then.'

When the probe set at 6.10 p.m. that evening, Professor Lovell declared: 'This really is a most magnificent achievement on the part of the Americans. Scientifically, 90 per cent of the achievement has already occurred to-day.' The probe was measuring three things: the variation in magnetic field, in micrometeoric impact, and in ion-content, out in space for the first time.

The work at Jodrell Bank has gone off without a hitch ('Tracking a radio star is a very much more sophisticated undertaking,' said Professor Lovell). During the week-end more than a hundred accurate fixes on the probe's position have been made (by deflecting the telescope off the probe's bearings and then tracking through it) and transmitted to Los Angeles for computing along with all the telemetred information from the probe's instruments.

The rocket failed to reach the moon because the velocity achieved at the burn-out of the fuel after launching was 34,000 ft/sec. instead of the required 35,250 ft/sec. In addition the launching angle was wrong by a few degrees and the probe would have missed the moon by 12,000 miles in any case. The payload contained a retro rocket which was to be fired in the vicinity of the moon in the hope that Pioneer would be put into an orbit around it.[1] When it was realized that the rocket was returning to earth, many attempts were made to fire this retro rocket by radio command so that it would be turned into a long period earth satellite, but the mechanism did not respond and in the early hours of Monday morning Pioneer burnt up as it re-entered the earth's atmosphere over the South Pacific.

[1] In fact 8 years were to elapse before this feat was accomplished. In 1966 the Russians achieved a lunar orbiter with Luna 10 on 3 April and the Americans with Lunar Orbiter One on 14 August. The Russians placed Luna 11 in lunar orbit on 27 Aug., and Luna 12 on 25 Oct. The Americans succeeded again with Lunar Orbiter Two on 10 Nov.

Although Pioneer failed as a moon rocket it was a striking scientific success; in particular it produced the first information obtained about the nature of the zones of trapped particles and the magnetic field out to nearly 80,000 miles from earth. For Jodrell Bank the consequences of that weekend eventually turned into our final solution. On the Monday morning the *Daily Telegraph* carried a message from its correspondent at Cape Canaveral:

> The Americans are relying almost entirely on British reports from Jodrell Bank for news of Pioneer and for the relay of the telemetry information coming to earth from its scientific packages. . . . Everywhere one hears generous praise for the scope and efficiency of the Jodrell Bank equipment. One American official said to me here today 'you British ought to shout out a bit more loudly about your achievements'.

33
The final appeal

At last we were drawing away from our days of misery and disgrace. At Jodrell we had seized our chances. Failure would have signalled our end. We succeeded; first with Sputnik (which seems in retrospect the most chancy and difficult experiment I've ever undertaken) and then with Pioneer. The University was now free to act, and for the first time fortune was with the Officers. Their release from the shackles of the Treasury and P.A.C. coincided with the enormous publicity in which the telescope had been applauded by every newspaper in the free world. As soon as the publicity of Pioneer began to fade my Reith Lectures were listened to by nearly a million people in Great Britain alone. In the third of these lectures I used the telescope to transmit a human voice to the moon and back. In his speech to the Court on 12 November (Chapter 31) Sir Raymond Streat had intimated that an appeal would be launched 'before the end of the month'. In the following weeks all preparations were made, many letters were written on an individual basis to industrialists and the Press were kept fully informed. It was officially launched on 28 November 1958 with the following explanatory leaflet.

The University of Manchester

RADIO TELESCOPE FUND

An explanatory Note for recipients of the Appeal

The University of Manchester is appealing for gifts, by way of seven or ten year covenants or single donations to a fund of £150,000, to provide that portion of the cost of the Radio Telescope which has not already been covered by grants from the Nuffield Foundation and the Department of Scientific and Industrial Research. Any balance in the fund will be used for developments in the instrument or ancillary equipment.

The aim is to free the instrument from debt and thus enable all available resources to be used in furthering the research work for which the telescope is designed. There is real urgency for this. For the next two or three years Britain has the only instrument capable of such work in the world and the University is anxious to exploit this situation to the full. This note has been prepared confidentially for those to whom the appeal is being addressed to equip them with the facts about the Telescope and its finances.

The building of the Radio Telescope was first proposed in 1951. Convinced of the importance of the conception and inspired by a belief that British Science could lead the world in this field the Council of the University adopted the project. It was thought at that stage that the Telescope could be constructed for £400,000. The Nuffield Foundation and the Department of Scientific and Industrial Research each agreed to find £200,000. The work was put in hand.

Some time later, as a result of developments not foreseen with precision at the outset, which were of a kind often encountered in pioneering enterprises on this scale of magnitude and originality it was found that to meet the circumstances and complete the Telescope in time for the International Geophysical Year, an increased expenditure of approximately £260,000 would be involved. The Council of the University took the best possible external advice, scientific and technical, and with this before them decided to proceed with the work.

Subsequently, the Public Accounts Committee in the perfectly proper execution of their duty, began to discuss the increases in the grants from public funds. The newspapers reported parts of the proceedings of the Committee. At that time the University was not at liberty to deal, in equal publicity, with the details of the matter because of its commitments to other parties. This made it possible for the public to think the finances had in some way been handled inadvisedly. That was not so. It can safely be asserted that no comparable pioneer enterprise has proceeded from conception to fulfilment without at least some unexpected developments and there is no reason to be unnecessarily apologetic because there were some during the construction of the Telescope. Fortunately, the final outcome fully testifies that the University's decisions at various stages were well taken. The increase of £260,000 was accepted by the Department of Scientific and Industrial Research as necessary in all the circumstances, after the fullest investigation, and on that basis the Department made a grant of £130,000, leaving the University to find donors willing to cover the balance. The total cost of the Telescope is far below what will soon be paid by other countries for similar instruments and much less than it would have been if the Council of the University had failed to authorise the carrying on of the enterprise at the time of the unexpected developments. The instrument is eminently successful from a technical and scientific point of view, and its contributions to recent events, notably the launching of satellites by Russia and America, have brought credit on British Science.

All over the world, tribute has been freely paid to the vision and enterprise which has resulted in Britain being first in the field with an instrument of this size. Prestige of this sort is a tangible asset in international trade and the University hopes this will be regarded as a further justification for donations from industry, finance and commerce.

The Bursar will be pleased to provide full information about alternative forms of covenant and taxation aspects.

Address: The Bursar, The University, Manchester 13.

Without exception the Press comments were now almost violently pro-telescope. The *Observer* anticipated the appeal on 16 November and commented: 'But fundamental science in Britain has come to a sad pass when a University is reduced to sending round the hat in this way to pay for a unique research tool that is appearing on hoardings all over the country as a symbol of British pride and industrial prowess.' This attitude permeated all comment. Sir Robert Renwick[1] wrote to *The Times*: 'the radio telescope has now become the envy of the scientific world. It is, therefore, to be hoped that one result . . . will be a clear demonstration that the telescope is a national asset of high importance and that Manchester University will immediately be removed from its present financial embarrassment.' The *Sunday Express* had a column headed 'THIS IS CRAZY. Of all Britain's immense technical achievements since the war, there is one today that remains unique . . . That enormous apparatus is one of the scientific wonders of the age. But there is one aspect of Jodrell Bank which can bring not pride, but anger.'

On 8 December at Jodrell Bank a Press conference was convened with Sir Raymond Streat, the Vice-Chancellor, and the Bursar present. At that moment the appeal fund stood at £23,000; mainly composed of the £10,000 donations from Henry Simon and Simon Carves previously promised (see Chapter 31). There was an avalanche of frustrated horror—mainly directed against the Government—in the Press. There were columns under such headings as 'Fantastic', 'Science on the H.P.', 'This is a bitter joke', etc.

It must be emphasized that in the University we never at any time took issue with the Government for not clearing this debt. The original 1952 grant was for 50 per cent. of the cost and when the final £130,000 grant was promised in the early autumn of 1958, the Government, through the D.S.I.R., had entirely fulfilled its original promise. However, the Government was attacked on our behalf both in the House and in private correspondence. It may well have found it politically expedient to have settled the final debt at that time, but as we knew, this was impossible because of the promises given to the P.A.C. which was an 'all party' body. There was a sharp exchange at question time in the House on 18 December. The Parliamentary Secretary to the Ministry of Works (Mr. Harmar Nicholls) made much of the new grant of

[1] Sir Robert Renwick Bt., K.B.E. (created Baron Renwick 1964). I had much to do with Renwick during the war when he was simultaneously Controller of Communications in the Air Ministry and Controller of Communications Equipment in the Ministry of Aircraft Production. Subsequently he always maintained a close and friendly interest in my affairs.

£15,000 which was about to be given to us; but of course this was for research and nothing to do with the debt.

The *New York Times* (10 December 1958) reported me as saying that the Americans regarded it 'almost as a joke that Britain should have produced an instrument like this and that its use is being restricted on financial grounds'. The British papers of the same date also quoted the beginning of this comment 'that the Russians regarded the telescope as a monument of this age and could not believe there was the slightest financial restriction'.

American dollars

The success of the telescope, particularly in the part which it played with the Pioneer space probe, fired the imagination of many Americans —the *Daily Telegraph* comment mentioned at the end of Chapter 32 was typical of the favourable attitude towards us which had developed across the Atlantic. For many years I had received grants from the U.S. Air Force and the Office of Naval Research for the support of our researches in radio astronomy. Indeed, during the period of our frozen relations with the Treasury when it was not possible for me to approach the D.S.I.R. for research grants, I depended heavily on these American sources for the continuation of the Jodrell Bank research programmes. However, these grants had nothing to do with moon rockets, and it must be remembered that I entered into the agreement with Col. L (Chapter 32) on condition that we would be free of the commitments and complications which might result from payment for this work.

Now the situation was developing rapidly. The personal and technical success of our cooperation over Pioneer delighted all concerned. The small American party who accompanied the trailer were immediately absorbed into our community. They brought over their families, who to my personal amazement seemed delighted to exchange the sunshine of California for the warm hospitality of Cheshire, in spite of the climate.

In this developing situation as we were preparing for the October event, I was asked to receive Congressman James G. Fulton the Republican representative of Pennsylvania, who was a key figure in the House of Representatives Space Committee, together with Dr. Charles Sheldon the Committee's assistant staff director.[1] They were greatly impressed not only by our financial plight which they thought was

[1] Apart from his bewilderment at our financial plight Mr. Fulton found it inconceivable that we were not constantly visited and encouraged by members of our own Government. It is a remarkable fact that apart from informal visits by our local

absurd but also by our cooperation with the American space effort.[1] On 15 October 1958 the *Daily Telegraph* carried a dispatch from their Washington Correspondent which began 'Members of the House of Representatives Space Committee, impressed by the contribution the Jodrell Bank radio telescope is making towards the lunar probe and satellite tracking, are ready to recommend next session that some American financial assistance might be arranged for Prof. Lovell's observatory.' The next day the *Manchester Guardian*'s Washington Correspondent reported that 'Staff assistants to a Congress Space Committee said today that they favoured a United States contribution . . . It remained to be seen whether the Space Committee members—members of the House of Representatives—would recommend such a contribution when Congress resumed in January. They said it was also possible that the National Aeronautics and Space Administration which has overall charge of non-military space projects . . . would propose U.S. support for Jodrell Bank.' Other newspapers were pointedly irritated at the necessity of non-British support for our work. The *Daily Express*, for example, carried headlines (16 October) 'U.S. to Buy into our Space Eye'.

The eventual result of these movements was that the Bursar negotiated an arrangement with the National Aeronautics and Space Administration (who took over responsibility from the U.S.A.F. for these lunar probes) that we would be paid a certain sum of money per hour—with a daily limit—when the telescope was engaged in these cooperative programmes with the American space agency. The Americans thought they were getting our services at a cut price; but

---

M.P.s (Col. [now Sir Walter] Bromley-Davenport and Air Commodore [now Sir Arthur] Vere Harvey) we had been totally neglected by our own Government. This situation was happily remedied on 28 Nov. 1958 when the Minister of Supply, Mr. Aubrey Jones, honoured us with his presence. I remember the Vice-Chancellor remarking to him at lunch time that his visit was an unusual encouragement and particular pleasure for us. If Mr. Aubrey Jones ever reads this footnote he may like to know that his visit that day was a memorable event for us, far transcending in its implications any encouragement which he might have expected to give by a visit in the normal course of his official duties.

[1] Mr. Fulton's visit was the first of many from members of the U.S. Congress, particularly those associated with the Committee on Astronautics and Space Exploration. Subsequently it was a great pleasure for me to speak before the Committee in Washington on 22 March 1962. The Honorable George P. Miller of California was in the Chair and I was surprised to find that his committee contained so many people familiar to me from their visits to Jodrell Bank. (The minutes of this meeting appear as the 'Panel on Science and Technology, Fourth Meeting in Hearings before the Committee on Science and Astronautics, U.S. House of Representatives, Eighty Second Congress, Second Session March 21 and 22, 1962 (No. 3)'. U.S. Government Printing Office, Washington 1962.)

the Bursar and I were delighted with this handsome arrangement which helped so significantly with our financial problem—particularly since in one of our graver moments a colleague had ironically said in a meeting that 'the thing would never earn a damned dollar'.

This arrangement meant that the Jodrell Bank telescope became an integral part of the U.S. 'deep space instrumentation facility'. The thrilling part which it played in the launching of the Pioneer V deep space probe in 1960 was the culmination of our good fortune and, as will be described in Chapter 35, brought our financial worries to a decisive end.

The British response to the appeal

Expectations that the successes of the telescope would make it easy to raise £150,000 turned out to be quite wrong. Although a considerable number of industrial firms responded to the appeal their magnanimity did not approach that of the Simon organizations. The attitude of the industrialists almost without exception was that the Government should clear the debt notwithstanding our repeated explanations that they had fulfilled their part of the original arrangement to bear half of the cost.

On three occasions during the spring and summer of 1959 we held large money-raising parties at Jodrell Bank, to which the industrialists and private people who had already contributed were asked and those we thought might still be persuaded to contribute. Hundreds of letters were written and all possible contacts explored. In spite of these efforts, which personally I found increasingly distracting and exhausting, the appeal fund had reached only £65,253 by July and had increased by a mere £530 by the end of the year. Apart from the two Simon contributions there were seventeen other gifts of £1,000 or over, and thirteen in the £500–£1,000 region.

We were tremendously grateful to the industrialists who responded, but they numbered less than a hundred and we were greatly disappointed at this limited response. On the other hand we were amazed at the response from private groups and individuals. The cynicism which had grown in us was drowned in a flood of letters from men and women, unknown to us, enclosing contributions varying from pennies to hundreds of pounds. The Manchester Branch of the Association of University Teachers circulated the staff suggesting contributions to an A.U.T. fund 'as a gesture of solidarity and confidence in the University'. Dozens of people sent contributions as a gesture of appreciation of my Reith Lectures (although of course there was no mention of the debt or the appeal in these lectures). School children sent in their pocket money and mothers wrote for children who were too young to write for themselves—

Dear Professor Lovell, The letter and P.O. are sent at the special request of my daughter P . . . (11½ years). She is vastly intrigued with everything and anything appertaining to the Heavens . . . She suggests that if every school child gave up 1 week's pocket money your troubles would be over.

The enclosed letter read

Dear Professor Lovell, Will you please accept this P.O. towards the cost of your equipment at Jodrell Bank. It is my pocket money for this week, but having read about your telescope I thought my money would be a little help towards it. I hope you get enough money to balance your budget.

Dear Sir, Form 4A has seen many pictures in the newspapers and magazines of your wonderful telescope, so hearing your appeal we thought we would save up and collect a small sum of money. On doing so we have managed to save fifteen shillings towards the telescope fund.

Dear Sir, My husband, daughter and self gave up our mid-week bar of chocolate for a few weeks to save this enclosed 10/–. It was not difficult and although it is not much it would be a great help to you I am sure if one tenth of the families in the country took it into their heads to send the same.

Dear Sir, Here is £1 towards your work at Jodrell Bank. I am an old age pensioner and this is in memory of a dear friend who would have loved to help.

Dear Bursar, Please accept this P.O. 2/– for the Jodrell Bank. It is a wonderful piece of technical and scientific mechanism and it is doing a marvellous job of work. I am so sorry but I cannot send a bigger amount for I am a disabled man.

The emotional nature of these private responses was almost overwhelming. We certainly did not lack public interest and goodwill but we still lacked money which we urgently needed. The figures quoted above show that after six months the impetus of the appeal had been exhausted. After a year, as 1959 drew to a close, the fund was growing at a negligible rate, and yet we were £65,000 short of the amount needed to clear the debt of £130,000 and £85,000 short of the sum of £150,000 for which we appealed.

A bizarre visit—Miss Sayaha alias Lady de Montford

Easter 1959, when we were still eager with anticipation that significant sums would come to lighten our burden, was marked by a strange visitor. Some time previously the Bursar phoned me to say that he had received an inquiry from a well-known London Business Consultant asking if he could bring a wealthy client to visit the telescope who wished to contribute to our appeal fund. Arrangements were made for the visit on Maundy Thursday (26 March 1959). Shortly before the the visit Bursar phoned again to say that the Business Consultant had thought it proper to let him know that his client was a lady of Indian extraction, Miss S. Sayaha, and that her name had to be kept secret.

On 26 March Rainford met them at Crewe and duly arrived at Jodrell Bank with the Business Consultant accompanied by a charming young lady. The luncheon and arrangements for the visit were the best we could produce and all went happily. Indeed Hanbury Brown who was born in India was able to talk at great length with Miss Sayaha about her bicycle manufacturing business in that country. At the end of the afternoon when we had done our best, Miss Sayaha had a brief whispered conversation with her Consultant, who then told us that she would donate £10,000 to the telescope appeal fund, and £10,000 to me to be used for whatever programme I wished.

I was electrified with joy. Not only a handsome contribution to the fund but a tremendously valuable gift to help us with our researches and other problems at Jodrell. I returned home bearing this wonderful news, which was dampened only slightly by a telephone call from Rainford on his way home after returning the visitors to Crewe station. He asked me what I thought of the day to which I made an ecstatic reply and said that I had already decided how to spend my £10,000. I asked Rainford why he had bothered to phone me to which he replied that he was uneasy and asked me not to say anything about the visit to the Chairman of Council until we had the cheque. He said he was uneasy because when I asked Miss Sayaha to sign the visitors' book at the end of the visit she had hesitated and had eventually signed Sahaya.[1] He thought it odd that the Consultant did not know how to spell her name (he had written Sayaha to Rainford).

This seemed to me quite a normal inconsistency (or even a typist's error) and I dismissed Rainford's doubts from my mind. However, the promised cheque was slow in coming. Rainford invited the Consultant to come again to see other parts of the University on 15 April because he had indicated that Miss Sahaya might be willing to contribute to other University enterprises. This visit did not materialize, but Rainford had lunch in London with Miss Sahaya and the Consultant on 28 April. A month had elapsed since their visit to Jodrell and they

[1] It is strange that although Rainford and I both remembered her with her pen poised hesitatingly over the visitors' book, but eventually signing, I cannot find her signature. The only signature for that date is that of the Business Consultant. Perhaps she eventually agreed to autograph some other document or book which I have since lost sight of.

promised then to send the cheque immediately. It did not appear. There were many letters and telephone calls in one of which it was said that the cheque had been posted but they understood there had been a fire in the post box and that it must have been amongst the letters destroyed.

At about this time the Business Consultant informed us that Miss Sahaya's real name was Lady de Montford and that it was no longer necessary to keep her identity secret. By this time I shared Rainford's suspicions that something was awry, since an extensive search failed to reveal the existence of any Lady de Montford. However, it was impossible to be absolutely certain about the affair because the Business Consultant was real enough and Rainford had visited him in his London offices. In November he informed Rainford that Lady de Montford was out of the country for a few weeks but would send the cheque 'as soon as possible'.

By that time even I had followed Rainford's doubts to the ultimate conclusion and I was full of admiration for his initial instinct and wariness. As far as I know not a single other person in the University knew that we had been the subject of a hoax in particularly bad taste, until our financial troubles were over.

Lady de Montford, alias Miss Sahaya, had completely vanished from my mind until one day a year later I noticed in *The Times*[1] a report of the trial and imprisonment of a London typist who had assumed the name of Lady de Montford. She was finally caught after arranging for the financing of an American tour by a combined Oxford and Cambridge Universities Rugby team which never took place. From the evidence at this trial it appeared that she had gone through with many escapades similar to the one at Jodrell Bank, and had also borrowed considerable sums from the Business Consultant. Everything she told us was a complete fabrication and she had never at any time been to India. Taking into consideration the fact that she had carried through several large-scale hoaxes of this character one feels it is a pity that the girl's talents could not have been better used.

I remain puzzled by the fact that a well-known London Business Consultant could be so unbusinesslike as to be trapped in this manner!

Before the end of the year we were greatly encouraged by a visit of another category. On 11 November 1959 we were honoured to receive a visit from H.R.H. Prince Philip, Duke of Edinburgh. He was

[1] 28 April 1960.

accompanied by the late Rt. Hon. Lord Woolton who was at that time Chancellor of the University. Prince Philip's great interest and understanding of our work created a memorable impression on all who were present.

The end of 1959—Still in debt

To the amounts subscribed could be added the dollars which we were beginning to earn as a partner in the U.S. deep space network. But the newspapers tell how things still stood at the end of 1959:

The Times December 3 1959.

£60,000 DEFICIT ON TELESCOPE

Jodrell Bank Anxieties

Sir Raymond Streat, chairman of the University Council, disclosed today at the annual meeting of the University Court that arrangements had been made for certain users to pay something against capital cost . . . the deficit, he said, was a cause of much anxiety but there was hope of further contributions from those who admired the achievements and recognised the importance of the telescope . . . Sir Raymond Streat called the building of the telescope a historic adventure which was by no means at an end. The decisions taken were not regretted although they had caused a deficit which might take some time to expunge.

In a leader on 3 December 1959 headed STILL IN DEBT the *Guardian* wrote:

The creation of this superb instrument has given Britain a place in space research that no other country yet has: we may have no satellites, but we can tell the Russians and the Americans just what is happening to theirs. It is a shameful fact that the Jodrell Bank telescope is still in debt and that the prestige it achieves for Britain still depends on borrowed money . . . The Americans (it must pass their comprehension that an immensely successful scientific enterprise is still begging for money) have offered to pay £50 an hour for the telescope's time when it is tracking their satellites. . . . So far the Government has refused to give any more help on the grounds that to give extra money to research projects that exceed their estimates would encourage extravagance. But the Jodrell Bank telescope is unique—it cost more than was estimated originally because no one had ever built such a telescope before. £60,000 would be a footnote in the nation's accounts, and much less than the Government makes the taxpayers find to pay for errors of judgement by its own departments over a year. The Treasury would be setting no precedent by enabling Jodrell Bank to start 1960 clear of debt: it would simply be making a deserved contribution to a project that has enriched the whole nation.

On 6 December the *Sunday Dispatch* carried large headlines:

THE SHAME OF JODRELL BANK

Schoolboys send pocket money to save our face

. . . The Chancellor of the Exchequer, Mr. Heathcoat Amory is to be asked in the House of Commons to save Britain the embarrassment of relying upon little school boys to uphold her prestige in the eyes of the world . . . Sir Arthur Vere Harvey M.P. for Macclesfield will urge him to pay off the debt . . . which prompted the President of the British Association Sir James Gray to say to me yesterday 'It makes me feel ashamed'.

However, Whitehall maintained a stony silence. The response from the appeal became negligible and as the spring of 1960 approached it seemed that the remaining debt would never disappear. The agitation on our behalf increased quite alarmingly. When the service estimates were revealed (including a huge multi-million sum for the proposed Fylingdales defence radar) the *Guardian* on 3 March carried a cartoon by Papas (reproduced here by kind permission of the Editor and Mr. Papas). On 15 March this paper had a column headed 'Stop being mean about Jodrell, Government told' which reported that Mr. Will Griffiths (the Labour

member for Manchester Exchange) had put down a question to the Minister for Science (Lord Hailsham). The *Guardian*'s Parliamentary Correspondent added 'there is no doubt at all that members on both sides are quite sufficiently impressed by the value and performance of the telescope to feel that Government meanness at this stage is out of place'.

On 25 March the *Guardian* returned to the subject once more:

The Government's position on the financing of the Jodrell Bank telescope was reaffirmed yesterday in the Commons by Sir David Eccles on behalf of Lord Hailsham, the Minister for Science. In reply to Mr. William Griffiths and other M.P.s he said that the contribution which the Government had already made to the capital cost of the telescope 'on terms which were well understood' was sufficient and that the rest of the money it was understood, would be raised by the University of Manchester . . . This situation appears to be accepted with resignation and fortitude at Manchester. . . .

The *Guardian*'s leader writer under the heading MEAN wrote on the same day:

The Government persists in its mean refusal to give any further help to Manchester University . . . Sir David seemed to think that the fact that 'only £60,000 remains to be collected' is in some sense a mark of the Government's past generosity . . . Besides if the Government thinks the sum so slender, could it not show some belated common sense and free the telescope from debt? The excuse to which Ministers repeatedly cling when challenged —that expenditure in excess of estimates ought not to be considered—ranks as sheer hypocrisy from a Government which has just presented the country with a bill for an additional £100 million in the current year . . . only a few weeks ago we were told that the Seaslug missile has already cost £40 millions to develop, instead of the £1·5 millions originally estimated, and that the bill for the Thunderbird missile has risen from an expected £2·5 millions to £27 million. Jodrell Bank has already brought more credit to British science and industry than the entire missile programme. Is it not time that the Government honoured its bills?

The *Guardian*'s comment that the situation appeared to have been accepted with 'resignation and fortitude' was indeed the case. Unfortunately Lord Hailsham who was then Minister for Science believed that we were responsible for the continual agitation. He did not realize that we constantly discouraged M.P.s from raising the matter in the House. Lord Hailsham came to Jodrell Bank on 28 March 1960, three days after the *Guardian* article. His visit had a most unfortunate beginning. The Rolls-Royce car in which he was driving to Jodrell with the Vice-Chancellor caught on fire at the Chelford roundabout three miles north of the telescope. He arrived late in a Press car which was following. He said that his desk was piled with letters every day from members of the public demanding that the Government should clear the debt and he was evidently disturbed at the political consequences of the mounting pressure on this issue. Only when he received repeated assurances from me and the Vice-Chancellor that we were not behind this agitation but, on the contrary, did what we could to damp it down, did he begin to enjoy his visit.

We were indeed at that moment resigned with fortitude to our debt, since the response to the appeal had effectively ceased. It seemed to us then that for as long as could be seen the telescope, which had become a symbol of British scientific and engineering achievement, would be burdened with a heavy debt, depressing to the spirit of those of us who lived with it, and acting like a millstone on our scientific researches for which we were urgently in need of money.

Then, as will be related in Chapter 35, unexpected and unpremeditated, a single phone call brought our troubles to a fairy-tale ending.

34

13 September 1959
A Soviet rocket hits the moon

DURING 1958 the United States made four attempts to send a rocket to the moon. The first two events have been described in Chapter 32: the first launched in August exploded less than eighty seconds after lift off; the second (Pioneer I) launched on 11 October, travelled nearly 80,000 miles into space before falling back to earth; the third (Pioneer II) launched on 8 November travelled only 1,000 miles. The fourth (Pioneer III) was launched by the United States Army and travelled 66,000 miles before falling back to earth. When I was in the Soviet Union in August both the intention to launch and the subsequent failure of the first American moon rocket were known. It was therefore natural that I should ask the space scientists the key question as to whether the Soviet Union was likely to follow the U.S.A. in an attempt to send a rocket to the moon. The trend of the answers, including a response from at least one scientist of undisputed authority who certainly could not have been in ignorance of the true situation, was always the same, namely that no attempt at a lunar rocket would be made until they could do the job properly and guide the rockets precisely. In November, after the failure of three American attempts, two eminent Russian space scientists visited Jodrell Bank and the trend of their answers was similar.

My reflections on these conversations coupled with the fact that as far as I could see there was no equipment in Russia capable of tracking a lunar rocket[1] led me to the conviction that the Russians had no

[1] The Russians had no tracking radio telescope of any significant size until late 1960. In a phenomenally short time they then constructed the radio telescope in the Crimea which I was shown during a visit to their deep space tracking station in the summer of 1963. This had an equivalent aperture of about 140 ft. (compared with the 250 ft. of Jodrell Bank) and was rushed through to be ready for the launching of their probe to Venus on 12 Feb. 1961. I asked how it could possibly have been built in a matter of months to which the reply was simply 'It had to be'. Previous to that the Russians must have used much smaller instruments for tracking but of course by the standards of those years they had unexpectedly powerful transmitters in the Luniks.

immediate intention of carrying out lunar or space probe experiments but were more likely to concentrate on the problems of getting a human being into orbit around the earth.

It was an erroneous judgement. In the late evening of 2 January 1959, when I was peacefully in my home following the score of Bach's Chromatic Fantasy and Fugue, the strident note of the Jodrell telephone extension confused the opening subject of the fugue. It was the controller informing me that the Russians had launched a rocket to the moon. That telephone call conveyed information of intentions which were in due time to disturb much more in my life than a Bach fugue.

Although we attempted to locate Lunik I with the telescope we failed to do so. To this day we are not certain why we failed. The Russians announced the frequencies on which it was transmitting. Probably our failure arose from inexperience and perhaps also the Lunik transmissions were commanded from earth and in those early days we did not understand the necessity for a constant watch through endless hours of quiescence. Although Lunik I missed the moon by 4,600 miles[1] it was a tremendous event at that time and made headline news everywhere. On 5 January 1959 a long leading article in the *Guardian* began:

> There will be unstinted admiration, not to say astonishment, for what the Russians have done with their rocket to the moon. By all reasonable tests it has been an unqualified success . . . Quite properly the Americans were saying last autumn that they would be content if their rockets get within fifty thousand miles of the moon: such is the precision needed to aim a rocket through the earth's gravitational field . . . Now it is clear as the regularity of the sputnik launchings suggested, that the Russians are masters at the accurate guidance of rockets off the ground . . . As an experiment even more than as a way of influencing friends and the friends of enemies, the Russian rocket has been a resounding success.

However, Lunik I was merely a curtain-raiser. The Russians were soon to demonstrate in no uncertain manner that they were masters at the accurate guidance of rockets. As far as I was concerned the drama of Lunik II was played out in less than thirty-six hours, for the first six of which I played cricket. I had written to my Russian colleagues complaining at the lack of information about Lunik I which would have enabled us to track it, and pointing out that this was a bad policy because of the existing doubts in the West about their claims. I got no answer until Lunik II. Even then there was no forewarning. Shortly after lunchtime on Saturday 12 September 1959 when I was

[1] And thereby became the first artificial planet of the sun moving in an orbit around it with a period of 450 days.

already in my car with the engine running to go to the cricket match, I was summoned back by frantic waves from a member of the family. The telephones had rung in chorus. From Jodrell, the controller to say the Russians had launched a rocket which would 'reach the moon' on Sunday evening, and on the other phone the voice of a pressman to ask what we were going to do about it. The answer to the latter was that I was going to play cricket,[1] during which time I would consider the matter and that in any case it would be more than thirty hours before the rocket reached the moon. It must be remembered that I was still feeling somewhat sour and disillusioned over the events of Lunik I and I had no desire to waste further time by a useless diversion of the telescope from its planned astronomical programme.

At that moment there were two research programmes on the telescope. During the daytime until about 6 p.m. in the evening Dr. J. V. Evans[2] was carrying out a radar experiment. After that time throughout the night the telescope was being used by Dr. H. P. Palmer[2] and his group in an experiment for the measurement of the angular sizes of the radio sources. There was a great queue of researches waiting for the telescope and its diversion to space probe uses was not universally appreciated by my colleagues. Nevertheless before resuming my journey to the cricket match I phoned J. G. Davies to ask him if we could meet at Jodrell in the evening to review the situation, warned Palmer that he might be late starting his programme that evening, and asked the duty engineer to make certain preparations in case we decided to use the telescope on the Lunik.

With the cricket match over I returned to Jodrell in the evening to find the American contingent in a state of frenzy. They had been continuously harassed by voices from Washington the entire afternoon asking and demanding that they should attempt to locate the rocket. Unfortunately they could only reply to the effect that I was playing cricket and that nothing could be done until I returned. The head of the American team gave me a vivid description of the monologue from Washington when this information was transmitted. I was unmoved by this interest from across the Atlantic and remained in a casual frame of mind until I unlocked the office containing the Telex machine on the off chance that there might be some message from Moscow. There was!

[1] I was then the skipper of the neighbouring village team in Chelford.

[2] Reference has already been made to Dr. J. V. Evans in Chapter 30. Unfortunately in 1960 he decided to leave us for America. Dr. H. P. Palmer came to Jodrell in 1952 and since that time has been in charge of the small group responsible for the important series of measurements on the angular sizes of the distant radio sources. He was elected to a Readership in Radio Astronomy in 1967.

Precise details of the frequencies of the transmitters in the Lunik and the exact coordinates calculated for the latitude and longitude of Jodrell Bank giving a time of lunar impact at 10 p.m. B.S.T. on the following evening. Since the telegram had been despatched from Moscow less than one hour after launching it was obvious that the Russians had prepared the calculations for Jodrell Bank in advance and that they had an almost audacious confidence in the success of the launching.

We sprang to action. Fortunately the aerial on the telescope being used in the research programme covered the Lunik frequency band. Palmer, who had arrived to start his research programme, helped to make a temporary diversion of his apparatus. To save time we carried a portable communications receiver to the swinging laboratory and suspended underneath the bowl we tuned in without difficulty to the bleep of the Lunik and found it to be on the precise coordinates given in the Moscow telegram, and at that moment 100,000 miles from earth. Meanwhile J. G. Davies, who had already arrived on the scene, located the signals on the other beacon frequency in the 19 Mc/s band. The measurements which he was to make with this apparatus turned out to be of cardinal importance. After spending some time in this fashion we returned the telescope to the zenith so that Palmer and I could descend to ground level and after further cabling alterations we were receiving the Lunik signals on both frequencies in the control building before midnight.

It was evident that the rocket was on the correct course for lunar impact and as the moon rose over the Jodrell horizon on the Sunday afternoon the place was already thronged with Press and radio men. However representatives of the news media from all over the world concentrated in such a short time at Jodrell Bank remains a mystery to me. The Lunik's transmissions were received in good strength immediately after moonrise. I suppose that, left to ourselves, tension would have mounted as the hours passed. Surrounded by dozens of pressmen, who were as fascinated as we were, the situation became dramatic in the extreme. We had hooked up our own observing equipment with cables running from the control room to another laboratory room and the television and radio people had also run their cables through the corridor of the building so that the normally peaceful laboratory had a quite fantastic air on that Sunday evening.

About two hours before the predicted impact time there was a most urgent call from America which the American team pressed me to take. Their messages that we had received the signals from the Lunik on course for the moon were not being believed. When the spokesman from across the Atlantic eventually expressed similar disbelief to me I simply held up the telephone in front of the loudspeaker and transmitted the sound of the Lunik across the Atlantic to him.

The predicted impact time was 1 minute after 10 p.m. B.S.T. The excitement and the tension remains vivid in the memory as the time approached. At 10.01 the bleeps were still loud and clear, at 10.02 we began to think that it may have missed, but 23 seconds later the bleep ceased—the first man-made object had reached the moon!

The news monopolized the front pages of the newspapers. The banner headlines read 'Jodrell Bank captures drama of the Space Age. 'MAN HITS THE MOON' (*Daily Express*); 'The Mind Boggles says Prof. Lovell. It's a Bullseye.' '10.2.23 . . . world hears news from Jodrell Bank' (*Daily Mail*); 'RED ROCKET HITS THE MOON. Signals Stop—Jodrell Bank Drama' (*Daily Mirror*). The *Guardian*'s more sedate headlines (14 September 1959) read 'Russian Space Rocket Hits the Moon. The first news came from Jodrell Bank when Professor Lovell and his wife sat in a small laboratory listening. The signals stopped suddenly indicating that the capsule had landed on the moon. The time was 10 hr. 2 min. 23 sec.' Their own reporter described the scene.

Jodrell Bank, Sunday. The bleeps that had been sounding for three hours through the control building at Jodrell Bank came to a sudden stop at 10 hr 2 min 23 sec tonight. There was no flickering and no fading. One moment—as the journalists crowded round a doorway and the B.B.C. man crouched on the floor with a primitive looking field telephone—the whistling sounds were there; the next, there was just the crackling of atmospherics. For a brief space all breath was held. Did the fact that the rocket had arrived about 1½ minutes late in a journey of 36 hours mean that it had not hit the moon's surface but had instead entered into orbit around it? There were journalists maintaining their reputation for cynicism—to believe it. But with the B.B.C. voice breaking in with 'Hello London, we're ready to go ahead ten seconds from now' it was clear that Prof. Lovell, the director of the Jodrell Bank Experimental Station, did not. The microphone was held in front of him and, cautious but somehow confident he told his interviewer 'We're inclined to believe that the lunar probe must have hit the moon'. It was enough. The stampede for the telephones began, the photographers' fingers itched upon the triggers of their cameras . . . Then it became time for adjectives. Prof. Lovell called the achievement a brilliant demonstration of the advanced state of Russian science and technology. One was amazed and astounded, the imagination boggled, it was quite fantastic from every point of view.

All the national newspapers on that morning of 14 September carried their own stories of those hours at Jodrell Bank. Those that sought the personal touches for the interest of their readers found many items which had little to do with the Luniks but which today re-create the atmosphere of the occasion. The *Daily Mail* carried a minute-by-minute diary: '9.31 Prof. Lovell, shirt sleeves rolled up, went into the tiny kitchen for a glass of cold milk and two digestive biscuits. His fair haired daughter, Susan, a 20-year-old nurse was making cups of tea and coffee for 40 press and radio men. Mrs. Lovell was washing up. 9.40 Dr. John G. Davies, 35-year-old senior lecturer in radio astronomy and Professor Lovell's right-hand man, slackened his red tie in the warmth of the control room.' According to the *News Chronicle* at 10.00 'Prof. Lovell took a handkerchief from his pocket and mopped his brow' and when the bleep stopped I 'rushed from the control room exclaiming "It's there! They've hit the moon!" In his excitement he tripped over television and sound cables strewn over the floor ready to record this moment in history.' The *Daily Express* described the scene and particularly the sounds: 'At 9.3 they sounded like the noise that comes from a motor-car starter as it struggles to turn over the engine from a weak battery. At 9.5 they were still there, but instead of being a "blur-e-e-ep" they were a slowly descending purr-r-r rather like the noise of a gramophone record running down. But at 9.6 in the background of the atmospherics the old high-pitched "blurr-ee-e-ep" came back again, getting gradually more high-pitched with every second.'

At that stage in the history of space science and technology—only twenty-three months after the launching of the first Sputnik—the Lunik was, of course, a most impressive achievement. The event came on the eve of Mr. Khrushchev's visit to the United States and the political significance of the conjunction of these two events was widely noted. The Americans were once more trailing, and to my amazement there were signs of disbelief. Mr. Richard Nixon the Vice-President said 'None of us know that it is really on the moon'. Mr. Truman, former President of the U.S., said the Russian feat was 'a wonderful thing—if they did it'. One of our own radio astronomical colleagues in America even told the Press that 'it would be quite easy to have a clock mechanism installed in the rocket so that the signal could be shut off at about the time the Russians said it was to hit the moon'.[1]

Fortunately during the last hour of the rocket's flight J. G. Davies had measured the doppler shift as the rocket fell under the gravitational field of the moon from which it was possible to calculate the general

[1] We learnt much later that one radio telescope near the East Coast of the U.S.A. had succeeded in recording the last minute of the signals from Lunik II as the moon rose above the horizon. The scientists worked in an establishment with military associations and were not allowed to announce their success. I was informed that any such release would have been incompatible with the official American reserve on the success of Lunik II.

region of impact.[1] He was able to do this by comparing the frequency of the received signal on the 19 Mc/s equipment with a standard in the control room. It was an absolutely crucial measurement which left no shadow of doubt that impact had occurred. But for this I think that many Americans would have continued to doubt the reality of the achievement. Some weeks later the director of the American moon programme remarked to me that Davies' measurement was a flash of genius. It was the first time such measurements had been made; the technique was to be refined and used in all subsequent moon experiments to establish the motion of rockets under the gravitational field of the moon.

In the U.K. the *Guardian* of 15 September reported: 'The Foreign Office was asked . . . whether Mr. Selwyn Lloyd's comment on the rocket—"I don't think many people are terribly interested in it"—represented the British Government's official view. The official spokesman said it could certainly be taken as Mr. Lloyd's comment.'

Whatever the official view might have been the real significance of the Lunik was admirably described by *The Times* in a leading article 'Lessons of the Lunik' on 15 September:

In putting a space vehicle on the moon the Russians have provided the most complete, as well as the most dramatic, proof of the length they now hold in accuracy of launching and control. They had been good at this from the beginning . . . now they have hit the bull's eye, or at least scored an inner. If this was a pure feat of marksmanship—controlled entirely at the time of launching—even a bull's eye is an understatement . . . On any view, and for whatever purpose it may be used, the rocket, in Soviet hands, has become a precision instrument. If one now asks, why they should bother to hit the moon when to circle it with instruments would be more informative, then two different answers can be given. The first and more obvious is that it is a demonstration to the world of what Soviet technology can achieve—simple, stark and impressive . . . If in addition, the demonstration is as well timed as this one, that too is impressive.[2]

Lunik III

The Russians were soon to add even more to the amazement of the world and the discomfiture of the Americans. On 4 October, precisely on the second anniversary of Sputnik I, they launched Lunik III designed to photograph the hidden side of the moon. To us all, coming only three weeks after the launching of Lunik II, the event was a complete surprise. I was not even at Jodrell. The Americans had intended to launch a Pioneer to orbit the moon early in October and I had been

[1] The measurements are published in *Nature*, vol. 184, p. 501, 1959.
[2] i.e. on the eve of Khrushchev's visit to the U.S.

invited to be present at the launching. While I was on my way to America the rocket exploded on the launching pad. Nevertheless I continued my journey to Los Angeles to liaise with the headquarters and laboratories of the S.T.L. who were then in charge of the American programmes. On Saturday afternoon Dr. George Mueller (in charge of the Pioneer project and subsequently head of the Apollo programme) decided to take me to Disneyland. While we were experiencing the mock journey to the moon, provided as one of the entertainments in Disneyland, the Russians launched Lunik III. There seemed something grimly ironical about this association of events on the precise day scheduled for the launching of the ill-fated American lunar rocket. The Americans were still not exhibiting the resolve and the skill necessary to achieve parity with the Russians—neither did they until the great incentive of the Kennedy programme materialized.[1] As *The Times* wrote on 7 October: 'the Russians have moved with a steady and impressive dignity. They have preferred to do first and talk afterwards.'

It was the first time I had been away from Jodrell Bank during a major space event and I was naturally anxious. I need not have worried. I soon heard over the U.S. radio that Jodrell had picked up the Lunik's signals, and when I spoke to J. G. Davies over the phone I found that everything was in order and that Moscow had transmitted the same information about the Lunik's coordinates as they had done for Lunik II. By the time I returned to Jodrell on 6 October the Lunik had already photographed, processed, and transmitted to earth the historic first photographs of the hidden surface of the moon.

I was delighted to receive the following letter from Academician A. N. Nesmeyanov, President of the Soviet Academy of Sciences (dated Moscow 25 November 1959).

Dear Professor Lovell,

We learned with great pleasure about your observations of the Soviet cosmic rockets II and III. Your participation in observations of Soviet cosmic rockets marked the beginning of our successful cooperation in cosmic space investigations.

In accordance with the scientific program of the automatic interstellar station it was foreseen that the principal information should be transmitted during the first revolution of the station around the Earth. The main purpose

[1] By 1966 the Americans were lagging only months behind the Russians in the soft landing of photographic equipment on the moon and the injection of orbiting space craft around it. At this time of writing (early 1967) there seems to be a near parallelism in the developments in both countries aimed at placing a man on the moon before 1970.

of these investigations was to obtain photographs of the back side of the Moon and to transmit them to the Earth.

It was supposed that during further revolutions of the interstellar station around the Earth additional data would be obtained which would make possible more detailed investigations. However at present communication with the interstellar station has ceased and this is why we did not send any ephemeris data after the first revolution of the station.

Trajectory measurements and computations show that the automatic interstellar station will complete about 11 revolutions around the Earth and at the end of March or the beginning of April will enter the dense layers of the atmosphere. Our scientists are at present engaged in the further analysis of the obtained data. We would appreciate it greatly if you could send us data of your observations of Soviet cosmic rockets II and III.

Taking this opportunity I am sending you the publication of the U.S.S.R. Academy of Sciences 'First Photographs of the Back Side of the Moon'. Later the USSR Academy of Sciences intends to publish a book with a detailed description of the photographs transmitted from the interstellar station.

We hope that our collaboration with you will be still closer in the future.

Sincerely yours,
(signed) A. N. Nesmeyanov

We were happy to send copies of our recorded tapes and other data on Luniks II and III to the Academy. When Academician Nesmeyanov's successor as President, Academician M. V. Keldysh, visited Jodrell Bank with other members of the Academy on 17 February 1965 I presented him with a tape of the last few seconds of the bleep of Lunik II before lunar impact.

35

Pioneer V—the end of our debt

In the face of these great Soviet space successes the Americans were having a bad time. The Pioneer series with which we were associated (Chapter 32) had fallen on evil times. Even the partial success of Pioneers I and III launched in the autumn of 1958 had faded in the face of the Luniks of 1959. To add to their discomfiture there were two more failures in the autumn of 1959: the one referred to in the previous chapter which was to have been launched the same weekend as Lunik III but which exploded on the launching pad some days previously, and another launched in November which fell into the Atlantic because the second stage failed to ignite.

In Moscow in 1958 when the first Pioneer failed in a blaze of publicity the Russians asked me why the Americans advertised their failures. In America, when Lunik III succeeded after the Pioneer had blown itself to bits at Cape Canaveral an American remarked ruefully to me that it was only necessary for an announcement to be made of American intentions for the Russians to do it first.

Two extracts from the contemporary Press show well the mood in America at this juncture. After the success of Lunik II Alistair Cooke in his dispatch to the *Guardian* (15 September 1959) wrote:

New York Sept. 14. Between twilight and dawn Americans have had time to recover from the humiliating trauma that hit them shortly after five last night when most television and radio programmes were interrupted by the flash from Moscow and the inaudible thud on the Moon. . . . It is safe to say that some American rocket wizards, and many a loyal officer on the staff of the Pentagon, were on their knees at the end of the afternoon praying that the Russians wouldn't make it, for the lesson has been widely expounded that a successful shot of this kind would confirm the Russian superiority in guiding missiles and in developing rocket fuel of enormous thrust. In the result, the scientists and the newspapers are dumbly or bravely echoing the verdict of Prof. Lovell at Jodrell Bank that it all amounts to a brilliant demonstration of the advanced stage of Russian science and technology. . . . At the moment the experts and the people are lost in a confusion of wonder and apprehension. The effects on the American psyche can hardly be less than the wounding effect of the first Sputnik.

After Lunik III was launched three weeks later the leader writer in the *New York Times* (5 October 1959) wrote:

If this historic feat succeeds all humanity will applaud the scientists and technicians who made it possible. Two years ago it would still have seemed incredible that by this date a man-made object would have struck the moon's surface and another would be on its way to photograph the far side of the moon . . . The sheer growth of mankind's—particularly Soviet—capabilities in this field staggers the imagination. . . . it can hardly be questioned that the two latest Soviet rocket feats prove that in space exploration and rocket-power we still trail the Soviet Union by a wide lap. Since we Americans find it annoying to be second best in matters scientific, it can be expected there will be increasing public pressure for a more effective space program than now exists. Greater financial support may be part of what is needed. But anyone who has followed our space and rocket progress knows that there are other obstacles besides lack of money: lack of scientists and engineers of top quality, lack of imagination in setting policy, friction and rivalry among the armed services and between the military and civilian directors, ideological opposition to the kind of effective Government rocket research which is represented by Werner von Braun and his group, competitive secrecy among firms engaged in some of the different branches of this technology. It is time to remove these obstacles and frictions and put this nation's rich resources efficiently to work on the conquest of space.

It was our good fortune at Jodrell Bank to be closely associated with the first striking American success in deep space in the spring of 1960. Pioneer V had originally been conceived as a probe to Venus, but it soon became clear after the failures of the lunar Pioneers that it could not be ready in time for launching during the suitable 'window' when the energy conditions for a launching to Venus were a minimum. The idea of Pioneer V therefore became a probe into deep space without a specific planetary objective. The spacecraft was built with a number of major scientific experiments in view. These included an apparatus designed by the University of Chicago for the measurement of cosmic rays in space over a wide energy range, apparatus from the University of Minnesota to measure the ionization and electric charge in a given region of space, a sensitive magnetometer from the Space Technology Laboratories in Los Angeles (who had overall responsibility for the project) to measure magnetic fields far out in space, and a micrometeorite detection device from N.A.S.A. If data from these experiments could be obtained as the probe travelled millions of miles into space then there was every prospect of rich scientific dividends.

Pioneer V was a brilliant success—and unexpectedly it brought our financial troubles to an end. For months prior to the launching we had been involved in unusual preparations. Whereas in all our previous work with space vehicles we had acted in a 'passive' capacity, that is we had received the signals transmitted by the probe, on this occasion we were to transmit to the probe as well in order to command its various functions. The plain fact was that the Jodrell telescope was still the only instrument which had any hope of transmitting with enough strength to the probe over distances of tens of millions of miles to command its functions. Our entire aerial system was therefore re-designed and a powerful transmitter was hoisted into the base of the bowl. This could only be done by cutting away a part of the membrane and was a major operation.[1] More trailers of equipment were flown over from Los Angeles and the American contingent working with us was reinforced.

11 March 1960 was an epic day in the history of Jodrell Bank. In addition to the command of Pioneer V's transmitters in space the telescope had the vital job of transmitting the signal to the probe to release it from its carrier rocket after it had been launched from Cape Canaveral. The launch was originally planned for 10 March. The count-down went to 70 seconds and was then called off because of fuel trouble. The next day there was no hold and a few seconds after 1 p.m. B.S.T. the rocket left its launching pad at Cape Canaveral. Twelve minutes later the probe came over the Jodrell horizon and its signals were immediately acquired by the telescope. At 1.25 p.m. when Pioneer was 5,000 miles from earth a touch on a button in the trailer at Jodrell transmitted a signal to the probe which fused the explosive bolts holding the payload to the carrier rocket. Immediately the nature of the received signals changed and we knew that Pioneer V was free, on course and transmitting as planned. For the rest of the day Pioneer responded to the commands of the telescope and when it sank below our horizon on that evening it was already 70,000 miles from earth. The next evening it was beyond the moon.

The signal transmitted by the telescope captured the imagination of the assembled pressmen. Everywhere on the Saturday morning, 12 March, there were headlines streaming across the page: 'Jodrell Bank sets satellite free. American Pioneer V in orbit around the sun' (*Guardian*); 'Planet control by Jodrell Bank. Telescope orders to U.S. Satellite' (*Telegraph*); 'U.S. launches Venus Satellite. Jodrell Bank sets off third stage' (*Times*); 'Jodrell Bank sends America's Beach Ball Satellite hurtling into orbit. Britain Calling Sun' (*Daily Express*); 'At 1.25 p.m. a tiny signal from Britain makes space science history. Jodrell's Big Hour' (*Daily Mail*), and so on.

[1] Eventually at the end of the experiments with the Pioneers, the transmitter was removed by helicopter.

Day after day the telescope commanded and recorded the vital information from Pioneer V as it sped into the depths of the solar system. A week after launching when Pioneer was 1,400,000 miles away H.R.H. Princess Margaret visited Jodrell Bank and pressed the button which transmitted the command signals from the telescope. In these early days the 5-watt transmitter in the probe was commanded. Pioneer also carried a 150-watt transmitter which was to be used when the probe's distance was so great that the 5-watt transmitter was insufficient. On 8 May, when Pioneer was 8,000,000 miles away, the commands to switch on this transmitter were sent out. It responded immediately[1] and we then had high hopes that we could command Pioneer to distances exceeding the original target of 50,000,000 miles. On 9 May the *Guardian* said 'A statement issued last night by N.A.S.A. said that Jodrell Bank commanded Pioneer V for the first time on the powerful 150 watt transmitter. . . . The statement says "This is truly an historic event" . . . Dr. Keith Glennan Head of N.A.S.A. said "To our British colleagues we extend our heartiest congratulations on their magnificent tracking and communication achievement".'

After another month there were signs of battery trouble in Pioneer and the use of the 150-watt transmitter had to be abandoned. Tracking and data acquisition continued until 26 June with the 5-watt transmitter. By that time the Pioneer's batteries were so weak that contact was lost at a distance of 22,462,000 miles.[2] It was under half the distance to which we had planned to work, but it was enough. The scientific results had exceeded all expectations. Favoured by a fortuitous series of spectacular solar events and by the perfect working of the scientific experiments on board, Pioneer V added greatly to our knowledge of the interplanetary space. Amongst other outstanding results the probe's magnetometer showed the existence of the disturbed magnetic field 50,000 miles from earth—the boundary between the earth's own field and the travelling fields ejected from the sun. The probe's cosmic ray equipment showed without ambiguity that the cosmic ray intensity decreased a million miles away from the earth when there was a solar disturbance simultaneously with the decrease observed on earth.

[1] That is after the 1½-minute interval while the signals were travelling to and from the probe.

[2] The silent Pioneer V is now in an orbit around the sun with a period of 312 days. It will remain in this orbit indefinitely. The carrier rocket from which it was separated by the signal from the telescope on 11 March is probably in a similar orbit. At the last point of contact on 26 June the signal from Pioneer was being picked up by the telescope four minutes after the command button was pressed, two minutes for the command signal to travel from earth to Pioneer and two minutes for the signal to return from the probe to earth.

Previously this phenomenon had been ascribed to an effect of the earth's local field. Pioneer showed that it was linked with changes in the interplanetary magnetic field associated with the plasmas travelling away from the sun when it was disturbed by sunspots and flares.

In July when Pioneer's batteries had died I was asked to go to America where I was honoured to be received in the White House by President Eisenhower who conveyed his appreciation of the part we had played in the accomplishment of Pioneer V's mission. I was much impressed by the President's detailed knowledge of these space activities. On 14 July 1960 the *New Scientist* carried a dispatch from John Lear, their American correspondent:

> Shortly after 4 o'clock last Thursday afternoon a music box the approximate size of an orange and the precise shape of the Man-made planet Pioneer V was presented as a gift from the American people to Dr. A. C. B. Lovell. . . . The box plays only one tune, but two sets of words go with it. Both sets are laden with patriotics. In this country the song is called *My Country, 'Tis of Thee*. Citizens of the British Commonwealth sing with equal fervour *God save our Gracious Queen*. Professor Lovell received this sentimental token from the hands of Dr. T. Keith Glennan, chief of the National Aeronautics and Space Administration, the agency responsible for orbiting *Pioneer V* between Earth and Venus. Without the faithful listening 250 ft ear of Prof. Lovell's radio telescope, the tiny artificial planet could not have been tracked through the sky and talked to across 22·5 million miles of space, and *Pioneer V*'s discoveries would have been unknown to Man. 'Nowhere' said Dr. Glennan, 'had NASA experienced such a truly fine example of international cooperation . . . so faithful, so efficient, so genuine, as with our good friends in Manchester England'. It was only proper that the man who 'conceived and fought for, helped raise the money for' the remarkable instrument of that cooperation should be recognised.

The Pioneer episode had been fraught with a terrible anxiety: shortly after launching the telescope cracked a bearing in the central pivot. It would have taken weeks to replace. There were urgent consultations with Husband, and with his blessing we continued day-by-day to track Pioneer with an explosive noise like machine gun fire occasionally emanating from the pivot of the telescope. Few people knew of this incident and it was a relief when, without any danger to a vital mission, we could remove the offending pivot and deal with the trouble.

Lord Nuffield

The response to the telescope appeal fund has been described in Chapter 33. In spite of the national agitations we still carried a heavy burden of debt when Pioneer V was launched. The money we earned in the American cooperative space programmes including Pioneer V helped to reduce this sum, but as Pioneer sped into space we were still shackled with a debt of £50,000; damping to our spirits, a constant source of anxiety and damaging to our progress.

Then suddenly, without warning, our burden vanished. A few days after Pioneer V was launched I answered my telephone. It was Lord Nuffield's private secretary, Mr. C. T. Kingerlee who had visited Jodrell Bank in 1955 with the director of the Nuffield Foundation. Lord Nuffield wished to speak to me: 'Is that Lovell?' 'Yes, my lord.' 'How much money is still owing on the telescope?' 'About £50,000.' 'Is that all? I want to pay it off.' I was almost speechless. I tried to thank him and to say that the telescope could never have been built but for the Nuffield Foundation. 'That's all right my boy, you haven't done too badly.'

It was a fairy-tale ending to the years of anxiety the depths of which were probably known only to my family.

Lord Nuffield asked that in recognition of the contributions made by his Foundation the telescope should be named the 'Nuffield–Lovell telescope'. There were too many difficulties and complications and eventually Lord Nuffield agreed that the whole establishment should bear his name. So we changed our name from 'The Jodrell Bank Experimental Station' to 'The Nuffield Radio Astronomy Laboratories, Jodrell Bank'.

The news of the clearance of the debt (a further £25,000 from the Foundation and a personal gift of £25,000 from Lord Nuffield) was given on 25 May 1960. Our relief was a major news item and the *Guardian* leader (26 May) under A DEBT OF DISHONOUR said:

> If ever there was a State investment of which a British Government could be proud it is the great radio-telescope at Jodrell Bank. It cost more than was estimated, because nobody in the world had tried to build such a telescope on such a scale before, and unexpected costs had to be met. The Treasury, although it had never been asked to pay the whole cost of the telescope because the Nuffield Foundation had contributed £200,000 shrugged off responsibility, simply leaving a debt of £140,000 to be paid by somebody else. A public appeal, as demeaning to a great nation as sending round the hat to make up a Cabinet Minister's salary, raised some £72,000, and the telescope earned further money for itself. . . . All this reduced the debt to £50,000 but still the Government made no move to help. Yesterday came the news that Lord Nuffield personally and the Nuffield Foundation have each contributed £25,000 to wipe the slate clean. It is a splendid gesture and we may take pride in the fact that there remain private citizens who are ready to shoulder Britain's debts. But for the Government this was a debt of dishonour, and remains so.

At last we were free of the shackles of the past and the present, released to think and plan for the future.

36

Jodrell Bank and the University

BEFORE World War II there was little 'big science' in the contemporary sense of that phrase. Particularly in universities the scientific projects were small and expenditures of a few thousand pounds were unusual. Indeed when a professor or member of staff moved from one university to another it was common for him to take his research equipment with him. The entire finance of science and of universities generally was on a microscopic scale compared with the position today. For example, the period between 1930 and the outbreak of the war can reasonably be regarded as one of the great phases of British science. Yet, in a typical year of that period we find that the *total* expenditure on research in the U.K. was only £6½ million of which £1½ million was spent on university research. In the financial year 1955–56, the total expenditure was £300 million and the amount spent on research by universities and technical colleges was £14·4 million.[1] Even when allowance is made for the threefold change in purchasing power the increase is remarkable.

This change was accompanied by the large increase in overall university expenditure associated initially with the expansion of the existing institutions and then with the creation of the new universities. The pre-war universities were in receipt of grants from the Treasury, but this income was generally small compared with that from private endowments and students' fees. In the University of Manchester, for example, the total income in 1938–39 was £300,000 of which about £100,000 came from the Treasury. In 1965–66 it was £6,200,000 of which £4,700,000 came from the Treasury.[2]

As this rapid growth occurred in the post-war phase the university grant continued to be a direct grant from the Exchequer channelled

[1] Council for Scientific Policy Report on Science Policy, May 1966, H.M.S.O. Cmd. 3007. The provisional figures for 1964–65 were £756·6 million and £55·9 million respectively.

[2] University of Manchester. Report of Council to the Court of Governors, Part I, Nov. 1966.

through the University Grants Committee and not through a Ministerial Department. University expenditure was not open to scrutiny by the Public Accounts Committee and this highly privileged position, although the envy of educationists in most of the rest of the world, came under an attack of increasing ferocity by certain sections of the community as the amount of the grant soared. This is the background to the complexities which developed over the telescope finances already described in Chapter 26.

The general attack on the favourable arrangements for British Universities was coupled with the argument that the universities were primarily teaching institutions and not places to engage in large-scale scientific research. This is not the place to deal with this fallacious argument. Unfortunately, throughout the years it has been, and remains, a source of trouble and discontent. Although the attack is mostly based from without it does, alas, have its supporters inside the universities, whose voices are heard with increasing prominence whenever financial stringencies occur.

Jodrell Bank was one of the first and certainly the most prominent of the large-scale post-war university excursions into great and expensive scientific projects. It will be obvious from the nature of the beginnings of Jodrell Bank described in the early chapters of this book that it was not initially conceived as a large-scale enterprise. When I first came with the trailers to Jodrell the only finance I had was the grant of a few hundred pounds which Blackett gave me from the Departmental funds. As we increased in size anxieties grew in certain quarters about the viability of running a university research enterprise twenty miles from the University centre, with the supposed consequent diversion of academic staff from teaching duties. The argument did not stand much investigation, because it was always a cardinal principle of the arrangement that no staff member working at Jodrell Bank should receive preferential treatment in respect of diminution of teaching duties. Nevertheless those who wished to attack the arrangement failed to take account of this fact. This attack was largely from within the University and was a most serious issue. If it had not been for the outlook of Blackett as head of the department, and of Sir John Stopford, the Vice-Chancellor, Jodrell Bank would not have gone beyond its trailers. It may surprise many people to learn that there never was an absolute victory. As we became larger and the telescope grew, the attack shifted its ground somewhat and demanded that we should be cast away altogether from the University. Constant vigilance is necessary even today.

The issue remained in the hands of Blackett and Stopford as long as the use of staff and funds was a small part of Blackett's department.

However, in 1954, when the University was preparing its estimates for the quinquennium 1957–62, it was obvious that Jodrell could no longer hide under the safe banner of the Physics Department allocation. The telescope finances were becoming a serious issue. Apart from that, my own research requirements had escalated. In 1954/55 I had nine academic staff working at Jodrell, the Technical and Administrative staff were costing £8,600 per annum, and the running and research expenses were £12,200 of which the University provided £4,700 and the D.S.I.R. £7,500. My submissions to the Vice-Chancellor on the increases necessary to continue the researches at Jodrell Bank when the telescope came into operation made it inevitable that the questions about the future organization for Jodrell Bank would have to receive more official sanctions than an internal University arrangement. Otherwise the Vice-Chancellor and the Officers would find it increasingly difficult to maintain their position about Jodrell in the face of the internal agitation, and the University Grants Committee had clearly reached the stage where a full understanding was necessary if it had to justify its finance of Jodrell Bank as part of a university.

The 'Ides of March'

I suffered the utmost alarm and anxiety that some manoeuvre would occur to separate Jodrell Bank from the University. Personally, apart from the war, I had spent my life in a university environment. The prospect of restrictions on choice of researches and freedom of action which another organization might entail filled me with dismay. I could not believe that Jodrell and the telescope could function in a restrictive environment. In addition to our other troubles in the early part of 1955 this fundamental issue arose in an acute form. In February I received a message from Sir John Stopford that Sir Keith Murray, the Chairman of the University Grants Committee, and Sir Edward Hale, the Secretary, would visit Jodrell Bank on 16 March. I wrote in my diary on 15 February 1955: 'after seeing the Vice-Chancellor . . . I am proposing to get out a memorandum on the future finance of Jodrell Bank and send it to Rainford before this meeting happens which will obviously be extremely critical for the future of Jodrell.' I did so; it was a straightforward account of our costs and demands for the future. But, I wrote on 28 February, this

'does not appear to have had the desired effect. . . . Rainford said this document would be the best way of damning my case . . . with dismay I heard from him the true purpose of the U.G.C. visit. It was to help them make up their minds whether projects such as Jodrell Bank, and Jodrell Bank in particular, should be the subject of their grants to the University . . .

found that the Vice-Chancellor was also unhappy about this document since it gave him official information of the very considerable financial demands, of which he would have preferred to have been unaware.'

On Rainford's advice I wrote to the Vice-Chancellor on 28 February asking if I might 'withdraw the memorandum. Under the circumstances I think it would be desirable for me to submit to you a document which would be suitable for transmission to the U.G.C.' I received his reply on 1 March: 'Dear Lovell, I herewith return the memorandum on the future of Jodrell Bank and as the Bursar would explain am greatly relieved at your action. Yours sincerely, John S. B. Stopford.'

I worked hard on another—a comprehensive document setting out our relations with the University, our teaching activities, emphasizing the importance of our postgraduate teaching and the ultimate disposition of our students, with Appendices on our research and extra-mural activities.[1] It was entirely academic, not a word of finance in it. First I took the precaution of showing the draft to Rainford and then I sent it to the Vice-Chancellor on 4 March: 'Subsequently I learnt from the Bursar that much more fundamental issues [than finance] were involved in this visit and this information has naturally caused me much anxiety. The Bursar encouraged me in the idea that it might be useful if I prepared a memorandum to refresh your memory of the relationship between Jodrell Bank and the University.' He replied: 'it will be a very useful resumé to prepare me for the catechizing to which I shall be subjected by the University Grants Committee. I will do my best for you.'

The arrangement was that the Vice-Chancellor would pick up the visitors at Crewe and have lunch on the way to Jodrell. Rainford came out nine days before the visit so that we could have a test lunch at a local pub and make sure that all the arrangements were in order.

There was no doubt whatever that our future depended on this visit and in spite of all our other troubles at that moment the visit

[1] It is a surprising fact that even today many people who are well in a position to know better fail to realize that Jodrell Bank is a research and teaching department of the University of Manchester. Nearly all the undergraduate teaching is done in the main University campus in Manchester but nearly all the postgraduate teaching takes place at Jodrell Bank. Even when this is appreciated we are often regarded as engaging in an 'ivory tower' activity with little association with practical needs. In fact half of radio astronomy is advanced technology. Recently (end of 1966) I had occasion to bring up to date the statistics of the posts taken by students who had carried out research at Jodrell Bank for the Ph.D. degree. Thirty-two per cent. had gone to British industry or Government Science—nearly twice the national average quoted for Physicists in Table V, p.12, of the 'Interim Report of the Working Group on Manpower Parameters for Scientific Growth' (the 'Swann Report'), H.M.S.O. Cmd. 3102, Oct. 1966.

monopolized my hopes and fears. In advance it became known as the Ides of March in the family circle. The day ended in joy not tragedy, and it remains my opinion that the negotiation of that dangerous passage was almost wholly the result of the Vice-Chancellor fulfilling his promise to 'do his best'.

Thursday March 17, 1955. The visit of Sir Keith Murray, the Chairman, and Sir Edward Hale, the Secretary of the U.G.C. occurred as planned yesterday afternoon with Rainford and the Vice-Chancellor. Rainford met me at the Power House at half past twelve and we drove to G where the party arrived in the Vice-Chancellor's Rolls from Crewe at about 1.15. We came to Jodrell at about 2.15 and after a fairly quick tour of the station there was a discussion in the library and the party left at 4.15. The nature of the visit was completely unexpected. Instead of the endless catechism about the part which we had to play in University life which I expected, I was surprised to find that the assumption right from the beginning was that the U.G.C. were going to finance this instrument. It was clear that my memorandum and the Vice-Chancellor's preliminary softening up in the drive from Crewe had worked miracles. In fact during the early part of the tour Rainford could be seen searching through his brief case for the original memorandum on finance which I had asked to be returned from the Vice-Chancellor's office. Hence he was well prepared for the financial questions.

Ever since the day in 1947 when Blackett brought the Vice-Chancellor to our trailers (Chapter 3), Sir John had seen Jodrell through the eyes of the distinguished scientist which he was, and not as a thorn in the flesh of a University administrator. The future of an institution was never more in the hands of a single individual than it was in his hands at that time. It would have been an easy administrative decision to have cast us off, pleasing and simplifying to an influential internal group. He chose the harder path because he believed in us and our future and because he knew that the work we were doing could only truly flourish in a University environment. I am happy that he lived long enough to see the realization of his judgement.

When Sir Keith Murray[1] retired from the Chairmanship of the University Grants Committee in 1963 I wrote to him and reminded him of that day he visited Jodrell. He was kind enough to reply: 'Jodrell Bank is really a jewel in the University crown . . . I will never forget the day that I came to see you with dear Stopford. You succeeded in firing the imagination of a dour Scot and I've never regretted my submission.'

[1] Sir Keith was created Baron Murray of Newhaven in 1964.

37

Envoi—into the depths of space

It has been my intention in this book to describe the building of the telescope and not to continue the story beyond the clearance of the debt in May 1960. After its completion in 1957 I have only related those aspects of Jodrell Bank which were so closely linked with its early years of work and which created the situation eventually leading to Lord Nuffield's phone call. The telescope became known, and is still widely regarded throughout the world, as a tracker of Luniks and space probes. Such operations have been subsidiary to its main use. As I write now early in 1967 the telescope has been in almost continuous use by night and day. More than 50,000 hours have been spent on various researches and only about 1 per cent. of this use has had anything to do with Luniks or space probes.

The telescope was conceived and built to study the universe and it is on researches into the depths of space that the instrument is used for 99 per cent. of the time. However, it attained world-wide publicity because of the Luniks, because of its use in the American deep space programme, and more recently (February 1966) because of the reception of the first photographs of the moon from the cameras in Luna 9 resting on the lunar surface. It may be that the telescope will be remembered in history for these activities for longer than it is remembered for its work in exploring the universe.

Jodrell Bank too will be remembered for the telescope, yet in the annals of scientific literature a future historian may remark that the first published scientific paper as a result of the use of the telescope (that on the observation of the carrier rocket of the Sputnik) is number 171 in the list. The first 170 had included descriptions of our work on meteors, on the moon, the ionosphere, the aurora borealis, the discovery of radio emission from the Andromeda nebula, surveys of radio emission from the Milky Way and many related topics. During that period new techniques had been evolved for the measurement of the angular size of sources of radio emission far away in the universe which were subsequently used with the telescope to form an important part of its research programme. With R. Q. Twiss, Hanbury Brown had evolved the principle of the stellar interferometer. Techniques derived for the radio spectrum had been reversed to apply to the optical spectrum and under the shadow of the radio telescope, through the hazy skies of Cheshire, they made the first measurement of the diameter of a visible star since the time of Michelson.[1]

As the telescope came into use after 1957 more and more of our research work became concentrated around it. By the time of the clearance of the debt to which this story goes we had published another eighty papers and apart from three publications on the Sputnik and one on the Lunik nearly all of these described the results of astronomical research programmes using the telescope. Today the Jodrell Bank publication list exceeds 350 and, although during the last two years another smaller telescope has been built (the Mark II), the majority of these describe the results of the researches with the original telescope.[2]

Now there is scarcely a moment in the day or night when on some recorder in a laboratory under the shadow of the telescope a pen moving on a chart cannot be seen. Its movements may record information about a planet in the solar system, more often about the nature of the hydrogen gas in the Milky Way and frequently about the emissions from peculiar galaxies so far away that the energy collected by the telescope started out on its journey through space many thousands of millions of years ago. We believe that in some of these cases, the 4,500 million years since the earth came into existence represent only a part of the time for which these radio waves have been on their journey towards us through space. In much of this work it is possible that the telescope is already penetrating as far into space and time as man ever will be able to penetrate. At a capitation cost which is already down to £10 per hour of use the telescope may, perhaps, be considered to have redeemed the agony of its construction.

Personally I find it hard to remember that the young men who are now studying at Jodrell Bank and using the telescope were only in the cradle when I first came here. In another two or three years the telescope will be as old as the young people who use it. In an age when science advances so rapidly that obsolescence occurs on the drawing board this seems to me to be a significant indication of the versatility of the telescope. It has survived a revolution in electronic techniques and it

[1] Hanbury Brown subsequently built a large version of this instrument which originally used two Army searchlight mirrors and took it to Australia where the sky conditions were good.

[2] An account of some of the contemporary researches and detailed descriptions of all the main radio telescopes at Jodrell Bank can be found in my small booklet *Our Present Knowledge of the Universe*, Manchester University Press 1967.

remains the largest fully steerable telescope in the world. One attempt was made to build a bigger one (by the U.S. Office of Naval Research in W. Virginia at Sugar Grove) but it was abandoned with only the bogie framework on site.

Apart from the bearing which cracked in the central pivot when the Pioneer V operation was in progress (see Chapter 35), and the replacement of some bearings in the bogies, the telescope has had only routine maintenance and painting although in continuous use for nearly ten years. It has worked through gales, snow, and torrential rain—surely a living tribute to Husband and his staff who transferred our vision to the drawing board, and the brilliant engineers who built it.

I see it now as I end this book, a most beautiful sight silhouetted against the evening sky.

38

Thirty years on

WHEN I wrote the previous chapter early in 1967 the telescope had been in use for nearly ten years. Fifty thousand hours of research work had been accomplished and the agonies of its construction, if not forgotten, had been redeemed. In the beginning I had argued with the critics and doubters that, if we could build the telescope, it would be a valuable scientific instrument for fifteen years. By the standards of obsolescence of major scientific instruments that was a long time and the few, who had enough faith to believe that the claim was not extravagant, doubted if such an unique and massive structure could remain in operation over such a time span.

At that moment in 1967 when the telescope was simultaneously penetrating into the depths of time and space and was still unique in its ability to track the Soviet and American space probes to the distant regions of the solar system, I was too occupied to think about the future. If I had been able to do so I very much doubt if I could have had the vision of what I see now—not ten, nor fifteen, nor twenty, but nearly thirty years after those extraordinary days when the telescope tracked the carrier rocket of the first Russian Sputnik. Remarkably, as I look up from this page I still see the great bowl and the filigree of the steelwork silhouetted against the setting sun. The window through which I see this beautiful sight is a different window from the one through which I looked in 1967. Engineers have apparently made the telescope ageless, but with the inevitabilty of the cycle of life the moment came in 1981 when I laid down the long-held burden of responsibility. However, it is my good fortune that I can still take a different direction along the corridor to the control room. I have done so now to ask the Controller on duty what object in the universe the telescope is being used to study this evening. His answer does not surprise me, but thirty years ago it would have been meaningless because the existence of this class of object in the universe was entirely unknown and unsuspected.

Indeed, the telescope has existed through an epoch where our knowledge of the universe has undergone a revolution paralleled only by the realization three and half centuries ago that the Earth was in motion around the Sun and was not stationary at the centre of the universe. The demands for observing time on the telescope are greater today than ever before and yet it is a most remarkable fact that the objects of investigation now were unknown when I first proposed the project. In Chapter 8 I wrote about that critical meeting in Edinburgh on 27 February 1950 when Appleton said that he was impressed by the range of problems that I had listed as capable of solution by a radio telescope of this size, but that he was "even more impressed by the possible uses of this instrument in fields of research which we cannot yet envisage". The words radio galaxies, quasars, pulsars, OH regions, and microwave background, for example, did not exist in 1950. If I had invented these terms and argued that the telescope would become an important instrument for research on such items I believe that even Appleton would have thought this was nonsense and my proposal would have been dismissed. Yet thirty-six years later nearly the entire use of the telescope is concerned with these objects. The details of this work have been published in over 800 scientific papers bearing the Jodrell Bank imprint and in "Out of the Zenith"[1] I have described the involvement of the telescope during the first thirteen years of its working life to 1970—years of excitement in astronomical discovery surely unparalleled in history. It is not possible to repeat these details here but I propose to summarise the development of some of the researches on radio galaxies and quasars for two reasons. First, because throughout the life of the telescope these researches have occupied more than a third of the telescope time and secondly because the desire to explore these objects in ever more detail led to the major developments that determined the nature of Jodrell Bank as it is now.

Radio galaxies and quasars

In Chapter 5 of this book I described briefly the circumstances of the discovery of radio waves from the universe and the immediate post-war work in England and Australia that led to the discovery of a number of localized or discrete sources of radio emission. The nature of these localized sources immediately became a topic of great importance and dispute. At the end of that chapter I mentioned that one of these radio sources (in Taurus) was related to the Crab Nebula (the remnants of the supernova explosion of 1054 A.D.). Beyond a further mention that at the time in September 1959, when a Soviet rocket hit the Moon (Chapter 34), H. P. Palmer was using the telescope for the measurement of the angular sizes of the radio sources, I did not give any further details of this work, although we were already deeply involved in these researches.

The story of the search for the identity of these radio sources is an intriguing episode in 20th-century astronomy. The erroneous interpretation of observational data coupled with the inherent obscurity of the universe typifies the tortuous path often pursued by scientists in the search for the underlying reality. At the time of which I write in 1950 the Cambridge radio astrono-

[1] Oxford University Press 1973, and Harper & Row 1973.

mers had published a catalogue of fifty of these radio sources in the northern hemisphere, and a further list of twenty-two in the southern hemisphere had been published by the group working in Sydney. Apart from the source in Taurus identified with the Crab Nebula, no one had succeeded in making any positive identifications with nebulae or other objects visible on the sky surveys of the optical astronomers. The Crab Nebula was in the local galaxy and the generally isotropic distribution of the other 71 radio sources led to the belief that they were also galactic objects in the solar vicinity. In fact, there was a widespread belief that they might be a class of hitherto undiscovered dark stars in the galaxy and were commonly referred to as radio stars.

In 1950 Hanbury Brown and Hazard used the Jodrell Bank transit telescope (figure 3 and Chapter 4) to study the radio emission from a strip of sky passing near the zenith at Jodrell. They searched the region around the extragalactic nebula M31 in Andromeda and made their historic discovery that this nebula, two million light years distant, was a radio emitter similar to the local galaxy. The signals were weak compared with those from the radio sources in the Cambridge and Sydney catalogues. These had been recorded by using small radio telescopes but it had required the far greater gain of the Jodrell transit telescope to reveal the weak signals from Andromeda. This led to the logical conclusion that the radio sources in the catalogues (the radio stars) could not be extragalactic because the only known radio emitting extragalactic nebula was a weak emitter of radio waves and there was no reason to suspect that the other extragalactic nebulae would not be similarly weak.

Thus, in 1950, this discovery at Jodrell Bank seemed to support the belief that the radio waves from the localised sources were emanating from a form of radio star in the local galaxy. There was a further observation in support of this belief. By this time the spectrum of the radio emission from the sky had been determined, and this differed from the spectrum to be expected from the free-free emission from the hot interstellar gas as originally proposed by Reber. The emission at long wavelengths was far too strong. Thus, the existence of radio stars in the galaxy seemed to be a solution of this difficulty with the excess emission at long wavelengths coming from these objects. Several of the finest astrophysicists of that era published a number of papers which gave a satisfactory explanation of the distribution and spectrum of the observed radio emission in terms of a combination of radio waves emitted by the radio stars and the interstellar gas. Indeed, at that time in the early 1950's there were few astronomers who did not believe that the 72 catalogued radio sources were dark radio stars in the galaxy.

During the next ten years a number of observational and theoretical problems were overcome so that by 1960 there had been a complete reversal of opinion about the nature and location of the radio sources. The early radio telescopes had very poor resolution and positional accuracies compared with the optical telescopes. All that was known about the angular size of the radio sources was that they were less than 8 minutes of arc in diameter. This upper limit is tens of thousands of times larger than the angular diameters of stars, hence at that time in the early 1950's we had no idea whether the radio waves were indeed originating in star-like objects or had their origin in diffuse regions of space far larger than that occupied by a star. As far as positional accuracies were concerned the "error boxes" in the assigned position of a radio source covered an area of the sky maps containing large numbers of stars and extragalactic nebulae. Thus, unless there was anything particularly unusual in the 'error box' (such as the Crab Nebula) even tentative identifications were suspect. In 1951 Graham Smith improved the Cambridge interferometer technique to such an extent that for the strongest radio sources he was able to determine their position with far greater accuracy than hitherto. In some cases the areas of the error boxes were reduced by 60 times. This improved accuracy enabled Walter Baade and Rudolph Minkowski to make a special search with the 200-in. optical telesocpe on Mt. Palomar for unusual objects in the position of the radio sources in Cassiopeia and Cygnus. By the early summer of 1952 their search had revealed that the Cassiopeia radio source coincided with a nebulosity in the galaxy having the characteristics of a supernova remnant. On the other hand the Cygnus source coincided with a peculiar extragalactic nebula estimated to be at a distance of about a thousand million light years. The unusual characteristics of this nebula suggested to Baade and Minkowski that there were two galaxies in collision.

The discovery of the extragalactic nature of the Cygnus source did not immediately change the opinion that the majority of the radio sources were in the local galaxy. Only three had been positively identified and two of these were in the local galaxy (Taurus and Cassiopeia). It is true that one major difficulty had been removed—that is, the weak emission from M31 in Andromeda could not be taken as an argument against the extragalactic concept—but the difficulty of explaining the galactic continuum emission at long wavelengths remained if the radio sources were not radio stars distributed in the local galaxy.

At that period in the early 1950's scientific contact between the Soviet Union and the West was poor, and until Soviet visitors spoke at an international symposium held at Jodrell Bank in the summer of 1955, little was known of the work of Soviet theorists on this problem. For the past four years a group of Soviet astrophysicists had developed the theory that the long wave emission in the galaxy was produced by high energy electrons moving in the interstellar magnetic field. Subsequent to this symposium the details of the Soviet work became better known and soon, this synchrotron radiation mechanism was generally accepted as the source of the long wave radio emission in the local galaxy. With this difficulty resolved a critical reason for the belief that the radio sources must be in the galaxy vanished. At the same period the Cambridge radio astronomers published a catalogue of 1936 radio sources and the number-intensity distribution of these led to the conclusion that they were a class of rare extragalactic objects similar to the source in Cygnus.

Thus, within a period of five or six years from 1950 there had been a complete reversal of opinion about the localised radio sources. At the beginning of the period they were believed to be dark radio stars in the local galaxy. By the end of the period the circumstantial evidence had led to the majority opinion that they were, on the contrary, a hitherto unknown type of distant extragalactic object to which the name 'radio galaxies' was given. I say circumstantial evidence because, even in 1958 at the Solvay Conference on the "Structure and Evolution of the Universe", in a list of 2000 localised radio sources there were available only 56 positive identifications. Only 7 of these were abnormal extragalactic objects of the Cygnus type. Amongst the others were 16 normal extragalactic nebulae and 23 in the local galaxy (20 gaseous nebulae and 3 supernova remnants). Thus, the direct observational evidence that the bulk of the radio sources were distant extragalactic radio galaxies had not by then materialised.

Two years later any remaining doubts disappeared when Minkowski identified one of the radio sources in Boötes with the most distant object yet discovered. This was the culmination of a sequence of measurements made at Jodrell Bank, first involving the transit telescope and then the 250-ft. steerable telescope when it became operational late in 1957. I have mentioned earlier that the problem of measuring the angular diameters of the localised radio sources was a major observational difficulty hindering their identification with objects known to the optical astronomers. In 'Out of the Zenith' I have described the sequence of long baseline interferometer measurements that steadily settled major issues about the angular sizes of the radio galaxies.

The problem of resolving power

The resolution of the unaided human eye is a few minutes of arc. With modern optical telescopes on high and dry mountain sites a resolving power of about 0.5 seconds of arc can be achieved. We see the Moon as an extended disc because it subtends an angle of half a degree. However, if the Moon were at the distance of the Sun it would subtend an angle of only 6 seconds of arc. With the eye we would see it as a point and we would need a good telescope to resolve the disc. If the Sun were at the distance of the nearest star it would subtend an angle of only three-thousandths of a second of arc and this is far beyond the resolving power of any telescope on Earth.

When we started to build the large radio telescope at Jodrell the achievement of a resolving power even approaching that of the human eye seemed a hopeless prospect. On a wavelength of a metre—the shortest wavelength for which we originally designed the telescope—the 250-ft. aperture paraboloid has a resolving power of only a little under one degree—twenty times inferior to that of the unaided eye. A single paraboloid of this type working in the metre waveband would have to be 140 miles in diameter to approach the resolution of the modern optical telescopes.

This fundamental problem of the comparatively poor resolving power of any feasible single radio telescope stimulated the radio astronomy groups in Sydney and Cambridge to the idea of the radio interferometer. Although it is not possible to make a single radio telescope with an aperture of 140 miles, for example, it is possible to separate two radio telescopes by that distance and so achieve the resolving power equivalent to that of the optical telescopes. There are severe technical problems in combining the signals from separated radio telescopes to provide the interferometric pattern that enables the diameter of the radio source to be measured. As long as the telescopes are close enough together so that the signals can be fed along a good cable to a common receiver the problem is easy—and it was this simple arrangement that led to the early measurements which could only assess an upper limit of several minutes of arc to the angular diameter of the first radio sources to be discovered. For the majority of radio sources the problem of separating the two aerials to such distances that they could be resolved turned out to be beyond the possibility of cable connections. It was this development using radio links to transmit the signals from a distant radio telescope to the telescope at Jodrell Bank that occupied a large part of our research effort from the early 1950's. We had the great advantage of the high gain of the transit telescope and in the early years of this research a smaller mobile aerial was moved to points at successively greater distances from Jodrell. By the autumn of 1955 all the radio sources in the zenithal strip which could be surveyed by tilting the mast of the transit telescope had been resolved except three. At that stage the mobile aerial was just under 13 km from Jodrell Bank. The radio frequency we were using was 160 MHz and at this spacing there had been no change in the amplitude of the interferometer fringes from these three sources. This implied that their angular sizes must be less than 24 seconds of arc.

We had been doubling the separation of the aerials for each successive series of measurements and another problem now arose because of the need to obtain a clear line of sight from the remote aerial to Jodrell. Fortunately, to the east the Cheshire plain soon begins to give way to the southern foothills of the Pennines. Those with good eyesight can just distinguish on a clear day a building on the ridge of these hills. It is the 'Cat & Fiddle', the highest inn in England. Henry Palmer, who was leading this small group, soon ingratiated himself with the landlord of this inn. Early in 1956 the linkage of the mobile aerial via a radio link was established from this inn to Jodrell and to our immense surprise these three radio sources were still unresolved which meant that their angular diameters were less than 12 seconds of arc.

The discovery of a remote radio galaxy

As Palmer pointed out in the published description of this work, this could mean only than the three unresolved sources must be of the same type as the remote radio galaxy in Cygnus. Rudolph Minkowski had been our main contact with the optical astronomers who were interested in this problem of the nature of the radio galaxies, and well before the results were published he realized the possible significance of identifying these unresolved radio sources. Using the 200-in. Palomar telescope he concentrated on one of these unresolved sources—catalogued several years previously at Cambridge as 3C 295 lying in the constellation of Boötes. The conclusion that the angular diameter was less than 12 seconds of arc meant that this radio source was 10 times smaller in angular extent than the Cygnus radio galaxy. Also it was 70 times fainter at the radio frequency of 160 MHz used in the Jodrell measurements. Thus, it seemed a logical conclusion that the object might be a radio galaxy far more distant than the one in Cygnus.

The first attempt which Minkowski made to photograph the object using the 200-in. telescope was unsuccessful because the position of the radio source was not known with sufficient accuracy. However, in 1959 the Cambridge and American radio astronomers determined a more precise position and this encouraged Minkowski to repeat the attempt at identification. He announced in 1960 that the photographs taken with the 200-in. showed the presence of a distant cluster of galaxies surrounding the position of the radio source. There were 60 galaxies in the range of visual magnitude from 21 to the limit of the 200-in. in an area of 3 minutes of arc in diameter in the centre of the plate. One of the brighter galaxies in this group, with visual magnitude 20.9, was precisely in the position of the radio source 3C 295. Further, since the spectrum of that galaxy resembled that of Cygnus in showing an unusually strong emission line, Minkowski had little hesitation in concluding that this was the object responsible for the radio emission. With the nebular spectrograph at the prime focus of the 200-in. telescope, Minkowski obtained two spectrograms of the galaxy, one with an exposure of 4.5 hours and another with an exposure of 9 hours. The measured redshift implied a recessional velocity of more than 40 per cent. of the velocity of light and a distance of about 4500 million light years.

This clear demonstration of the belief that the small-diameter radio sources might be objects analogous to Cygnus but at much greater distances was instantly realized to be of immense significance in cosmology. The predictions derived from the various evolutionary theories and those of the steady-state began to diverge significantly in the regions of space and time to which the telescopes had now penetrated by the identification of this 3C 295 radio source. In particular, the measurement of the angular diameters emerged as a possible decisive test between the steady-state cosmology then in vogue and the evolutionary cosmology of general relativity. At a symposium in Paris in 1958 Fred Hoyle had pressed that these measurements should be made. On the steady-state theory the apparent angular diameter of an object of given absolute size would decrease steadily as its distance in the universe increased. However, that apparently commonsense result does not apply for the evolutionary cosmologies of general relativity. The theory of general relativity makes the somewhat obscure prediction that, although the apparent angular diameter first decreases with increase of distance as expected, ultimately a minimum value is reached. The theory then predicts that at greater distances the apparent angular diameter of the object would increase. Eventually other factors intervened to reduce interest in this cosmological test. The discovery of the microwave background radiation in 1965 appeared to be decisively in favour of the evolutionary cosmologies. Also as the radio source measurements progressed, the great spread in size and the complexity of the structure of the radio sources militated against any such simple cosmological test.

Nevertheless at that time in the late 1950's the possibility of making this cosmological test provided a strong stimulus to proceed with the angular diameter measurements of the radio sources, with even longer separations of the interferometer aerials. This work soon led to a discovery of great significance.

The discovery of quasars

At this moment of intense interest in the angular diameter measurements we were able to use the Mark I telescope. Hitherto the use of the transit telescope had limited our investigations to the relatively small number of sources which could be detected in the zenithal strip. Now we had complete coverage of the northern sky. During the years when Jodrell Bank and the telescope became known to the public for its association with Earth satellites and space probes its major use was, in fact, for these angular diameter measurements. In the two years from the autumn of 1958, 37 per cent. of the telescope time was used on this work and when the initial series of measurements had been completed in 1961, 4000 hours of the telescope time had been absorbed in this work alone.

In the early stages with the Mark I telescope longer baselines were obtained by using the remote aerial at sites in North Wales but a number of sources remained unresolved. The remote aerial was then moved to an airfield in Lincolnshire, and the British Broadcasting Corporation allowed us to mount a repeater system on one of their television masts high on the Pennines. Even during the North Wales phase of this work results of great consequence were obtained. The stimulus for the identification of the radio source 3C 295 by Minkowski had been that the angular diameter was less than 12 seconds of arc. Now it was resolved and found to be 4½ seconds of arc in extent. At the same time, 7 out of the first 90 sources measured were found to have angular diameters less than 3 seconds of arc. The logical conclusion was that these must be radio galaxies of the same type as Cygnus and 3C 295 but at an even greater distance in the universe. The possibility of obtaining optical identifications immediately assumed great importance. Although the sources were listed in the Cambridge catalogue, the positional accuracies were too poor to enable a search to be made with the 200-in. telescope. Fortunately the group at Owens Valley in California were able to use their interferometer of two 90-ft. paraboloids to determine the position of one of these sources—3C 48—with six times greater precision than that given in the Cambridge catalogue, and the attempt at optical identification with the 200-in. Palomar telescope became possible.

The first news of this attempt at identification was given by Allan Sandage during the meetings of the American Astronomical Society held in New York City between 28-31 December 1960. In September he had used the 200-in. Palomar telescope to photograph the region of the sky in the position of this radio source, 3C 48. His announcement that a faint blue star existed in the precise location of 3C 48 and that it was accompanied by a faint wisp of nebulosity caused a sensation in the astronomical world. He was unable to identify any spectral lines that would have enabled him to measure the redshift and hence the distance, as Minkowski had done for 3C 295, and he concluded that this must be a relatively nearby star with most peculiar properties. Shortly afterwards Sandage identified two more of these small diameter radio sources with similar star-like objects of a blue colour. For nearly two years astronomers were startled by these discoveries. The smallest diameter radio sources yet measured, for which all logical reasoning had indicated that they must be remote radio galaxies in the universe, were, it seemed, apparently nearby stellar objects in the Milky Way.

The great difficulty was that the spectrum of these objects could not be identified and a veil of obscurity surrounded the subject. In 1962 observations by a former Jodrell Bank student, Cyril Hazard, using the radio telescope at Parkes in New South Wales, gave the key to the remarkable nature of these objects. In 1960 Hazard had used the Mark I telescope at Jodrell to study the variation in the signals from a radio source as it was occulted by the Moon. He was able to improve on the published position of this source by 20 times in right ascension and by 240 times in declination. He left us in 1961 to take an appointment in Sydney University, and in 1962 he had the chance to use the large radio telescope at Parkes. With this he employed the technique he had developed at Jodrell to study the occultation of one of the unresolved small diameter radio sources known as 3C 273. The accurate position thereby determined led to an identification with another peculiar star-like object accompanied by a jet or wisp, on a plate taken with the 200-in. telescope. The issue of *Nature* for 16 March 1963 contained details of this work and of a comment by Maarten Schmidt on the nature of the 'star'. He had studied the spectrum and had concluded that the only explanation was that this was not a star in the Milky Way but a hitherto unknown type of distant extragalactic object. The same issue of *Nature* contained the even more remarkable news that, stimulated by Schmidt's conclusion that the 'stars' were distant objects, Jesse Greenstein and T. A. Matthews had again studied the spectrum of 3C 48 and had concluded that the observed spectrum could be made to fit a red-shifted spectrum of hydrogen lines if 3C 48 had an apparent recessional velocity of 110,200 km/sec., implying that it was one of the most distant extragalactic objects known. The enigmatic objects which had been believed for over two years to be stars in the galaxy were, on the contrary, found to be amongst the most distant objects known in the universe.

The way was now open to further identifications, and by the end of 1963 nine of the small diameter radio sources, superficially star-like in appearance, had been identified as distant extragalactic objects of this hitherto unknown class. At that time the determination of the redshift of one of these sources—3C 147—had implied a distance of over 5 billion light years—more distant than the radio galaxy 3C 295 in Boötes. These strange objects, at first mistaken for stars in the Milky Way, transpired to be distant extragalactic constituents of the universe, generating immense amounts of energy in comparatively small volumes of space. Initially they were known as quasi-stellar radio sources, but soon the acronym quasars became generally accepted.

In the years that followed many more quasars were discovered and were recognised to be an important constituent of the Universe. Now, more than twenty years after the recognition that we were dealing with a hitherto unknown type of celestial object, about 10,000 extragalactic objects have been identified with radio sources. Some 30 per cent. of these are quasars and the majority of the remainder are radio galaxies.

The pursuit of long base-line interferometers

The early incentive ultimately leading to the discovery of quasars came from the Jodrell Bank measurements that revealed the existence of radio sources with angular diameters of less than a few seconds of arc. By 1961 the diameters of 384 sources had been measured using the Mark I linked to the remote aerial on the Lincolnshire airfield. Five of these sources were still unresolved so that their angular diameters must be less than 0.8 seconds of arc. I have mentioned that at this stage it was expected that the angular diameter measurements would lead to a decisive conclusion about the major cosmological issues. The discovery of the quasars now provided further incentives. Once more it seemed that the radio sources of smallest diameter could be used as the targets for the optical identification of even more distant quasars. To this was added the mystery of the great energies produced in the quasars. Even with the upper limits of diameters now set to the unresolved sources it seemed that these distant quasars must be generating energy equivalent to millions of suns in a small volume of space.

At the moment when these strange features of the quasars were emerging I went to the Soviet Union as the guest of the Academy. This generous invitation was arranged by the Academy to thank me for all that we had done to help them with the tracking of their space probes. It was the summer of 1963, and at the Crimean Astrophysical Observatory a summer school for young Soviet astronomers was in progress. The brilliant Soviet astrophysicist I. S. Shklovsky was there and after one of my lectures we had a long discussion during which he displayed great excitement about the physical nature of the quasars. Of the few that were unresolved he was convinced that they were immensely distant objects in the universe and he urged that we should extend our interferometers until we could discover their angular extent. To the purely cosmological case for improved resolving power Shklovsky had added the physical argument that entirely unknown processes of energy production might be operating in these remote regions of the universe.

When I returned to Jodrell I persuaded Palmer of the strength of Shklovsky's case. We cancelled the immediate plans for more detailed studies with the interferometer spacings already in use and Palmer set out to find a more distant site than Lincolnshire for the mobile aerial. We placed pins in the map of Yorkshire and Palmer soon found a site on a farm in Pocklington that would treble our resolving power. The repeater station was moved from the BBC television mast to another high mast, used by the Post Office, at a site appropriately known as Windy Hill. Less than a year after my talk with Shklovsky thirty-four sources had been measured—and five were not resolved, implying an angular extent of less than 0.3 seconds of arc. Two of these were soon identified optically and Shklovsky was right—they were quasars more distant than any previously found and presenting a whole range of problems about the mechanism of energy production.

In September 1964 it was my good fortune to be one of the small group of scientists in Brussels for the Solvay Conference on "The Structure and Evolution of Galaxies". I had been able to present these new measurements, which emphasized the problem of accounting for the production of large amounts of energy in small volumes of space. The Chairman of the Conference was J. R. Oppenheimer, and in introducing the final session he said:

> In this session we will discuss topics about which we know very little. Perhaps then we will all be treading on common ground. We are going to discuss possible sources of very large energies—which one detects, and the still larger energies which one infers from some of the greatest radio-emitters . . . The obvious candidates for great energy are nuclear energy and gravitational energy; neither has proved to be the answer to the question of origin of these great sources, nor have we got proof that these are not adequate. But simply, they do not work because of the energy amounts, the time scale of their release, and the forms of energy they release.

Once more I returned to Jodrell with the determination to set closer limits to the angular sizes of these unresolved quasars. There were many problems and not all of these were technical. Understandably, Palmer and his group, having spent such an arduous winter in establishing the Pocklington base, wanted to continue with that site. But this would not improve our knowledge of the five sources we already knew to be less than 0.3 seconds of arc in angular extent. Eventually we found the solution with an arrangement to use an 82-ft. radio telescope that had been built on the runway of the aerodrome I had used during the war at Defford in Worcestershire. The airfield was no longer operational, and this radio telescope was one of a pair which had been built for measurements of interest to the Defence department. The distance from Jodrell was almost the same as that to Pocklington, but the Defford telescope was larger than our mobile aerial and could be used on a wavelength of 20 cm instead of the 70 cm wavelength we had so far been using. This immediately improved our resolving power by three and a half times. By May 1965 the new arrangement was working and the five sources previously measured were found to have an angular extent of 0.1 seconds of arc or less. These were remarkable results. Clearly we were concerned with physical processes in these quasars emitting an energy equivalent to millions of suns, but contained within a spatial volume minute in comparison with galactic dimensions.

Another year and, with a still further reduction in the wavelength of the Defford-Jodrell system, another fifty radio sources had been measured. Now the upper limit to the angular extent of some of the quasars could be set at 0.05 seconds of arc. The distance of the nearest star to the Sun is 1.3 parsecs. The previous measurements had indicated that the spatial volume in these remote objects, in which such vast amounts of energy were generated, were minute compared with galactic dimensions. These new measurements showed that the spatial volumes involved were to be compared only with volumes of space containing a few of the nearer stars in the solar neighbourhood.

For more than a decade we had led the international field in the search for higher and higher resolving powers. When the construction of the Mark I radio telescope began the achievement of resolving powers in the radio spectrum equalling that of the unaided eye, was an unfulfilled dream. Now, in 1966, the telescope that had such a tortuous history had been instrumental in the attainment of resolving powers in the radio spectrum thirty thousand times superior to those of the early 1950's—not only better than the human eye, but ten times superior to the resolution of the largest optical telescope on Earth, operated under the best seeing conditions.

With these remarkable results we reached our climax. Others had been stimulated by this work to seek various technical solutions to the problem of multi-million wavelength baselines—and in this work we had neither the manpower nor financial resources to compete in their time scales. However, our use of the telescope at the Defford airfield—the property of the Defence department—eventually led to an operational system that once more achieved international status in the research on the angular structure of the radio sources. When this book was first published in 1968 neither the system nor the astronomical results were conceivable, and to conclude I must describe the sequence of events that led to the system as it now exists.

39

The Mark IA telescope

As soon as we were freed from the burden of our debt in 1960, encouraged by the success of the telescope, I made plans for a whole series of new radio telescopes. For the next twenty years I became involved in a number of political and financial battles that at times became uncomfortably reminiscent of the story related earlier in this book. When I retired in 1981 I attempted to unravel the complex series of interactions and in 1985 my account was published in *The Jodrell Bank Telescopes*.[1] It is ironical that my major failure to build a telescope of far greater dimensions undoubtedly saved the Mark I from decay.

During the years of research on the angular diameters described in Chapter 38 we had succeeded in building two more radio telescopes. One of these, known as the Mark II, was smaller than the Mark I but had a paraboloidal surface of greater accuracy to enable us to work on the shorter wavelengths which had been opened up by new developments in receiving equipment. This Mark II was built on the site of the 218-ft. transit telescope and came into use in 1964. A telescope of similar size and shape, the Mark III, was a less accurate and more flimsy structure. It was built specifically as a telescope to be used in association with the Mark I as an interferometer for the angular diameter measurements. We built it on farmland 24 km south of Jodrell in such a way that it could be dismantled and moved to other sites as the research programme warranted. This Mark III telescope came into use in 1966 and now, twenty years later, it is still on the original site so firmly integrated into the network of Jodrell telescopes that any idea of moving it elsewhere has long since been abandoned.

These Mark II and III telescopes were subsidiary to my major ambition to build a telescope of immense size. A rough sketch of this proposal made with Husband in 1960 reveals the concept of an elliptical paraboloid with a minor axis of 500 ft. and a major axis of 1,500-15,000 ft. This became identified as the Mark IV and caused so much consternation in the grant-giving bodies that a high level committee was established to reconsider the entire problem of developments in radio astronomy in the United Kingdom. During 1963, when the work of this Committee had destroyed the vision of this telescope we began serious design studies on a more conventional instrument, that is, a

[1] Oxford University Press 1985.

telescope of the same steerable type as the Mark I but with an aperture of at least 400 ft. instead of the 250-ft. bowl of the Mark I. By 1970, after years of study with Husband the design of this telescope was complete. A scale model had been built and in May 1970 a Press conference had been convened to announce the decision to proceed. As these arrangements were about to be promulgated the Labour Prime Minister (then Mr. Harold Wilson) decided to call an election and all major new scientific projects of the Mark V type were suspended. This election on 18 June 1970 led to a change of Government and another year elapsed before we were instructed to seek new prices for this telescope. The economic problems of the grant giving bodies, coupled with the impact of severe inflation, eventually forced us to reduce the size of the proposed Mark V to 375 ft. This became known as the Mark VA. In 1970 the estimated cost of the 400-ft. Mark V was £6 million. By 1974 the estimated price of the 375-ft. Mark VA had risen to over £20 million and it was then finally abandoned. The Public Accounts Committee once more decided to give its attention to our affairs and in August 1976 issued its report on the wastage of £750,000 on design fees for a telescope that was never built. The Public Accounts Committee congratulated the grant giving body (the Science Research Council) on its 'step-by-step' approach—the approach that had killed the project and had caused many of us at Jodrell to waste an immense amount of time over those years of design and discussion.

The Mark I to Mark IA conversion

That is a summary of the background against which a vital decision was taken in 1967 that led to the survival and transformation of the Mark I telescope. While the Mark II and III telescopes were being built, and the proposals for the Mark IV and then the Mark V were under discussion, I had devised a scheme for the modification of the Mark I. It will be remembered that when the idea of the telescope emerged around 1950 there was little interest in using wavelengths less than about one metre. Suitable electronic equipment for use on shorter wavelengths did not exist and there seemed no compelling astronomical reason for working at these short wavelengths. The discovery of the 21 cm line emission from neutral hydrogen changed this outlook and we were able to make modifications during the course of construction of the Mark I so that it could work on this wavelength, albeit with reduced efficiency (Chapter 17). By the early 1960's new types of electronic equipment became available and astronomical interest in the study of the radio emissions at short wavelengths grew rapidly.

In 1963 I sought and obtained a small grant from the DSIR to enable Husband to investigate a modification that would give us the necessary improved accuracy of the paraboloid. By the spring of 1964 he had produced his report in which he proposed to replace the single stabilising 'bicycle wheel' by two load bearing wheels running on an inner railway track. This would reduce the load on the trunnions and would lead to a firmer control of elevation. Partial introduction of computer control was proposed and there was discussion about changes to the reflecting membrane and the arrangements for holding the equipment at the focus. The estimate in April 1964 for this work was £268,000. I did nothing about this because any application for such a sum would certainly have jeopardised the Mark V concept then proceeding favourably. Apart from this financial conflict, we were at that time heavily involved with the Soviet and American space programmes as well as with the researches described in the previous chapter and it seemed a most inappropriate moment to put the Mark I out of action for such modifications.

I placed Husband's report in the files and, more than three years later, had almost forgotten its contents until a series of events just before Christmas 1967 caused me to take sudden action about modifying the Mark I. At that time I was Chairman of the Astronomy Space & Radio Board of the Science Research Council—the grant giving body to Universities which had taken over these responsibilities from the DSIR in 1965. The Board was faced with at least three major projects that would be impossible to finance within the scope of any feasible budget of the Council. These were the UK share of the 150-in. optical telescope in Australia, a radio astronomical project for Ryle in Cambridge and our own Mark V telescope. I had already pressed for priority to be given to the optical telescope[1] and now I offered to delay the Mark V project in favour of Ryle's new system. I have no doubt that in the interests of astronomy at that time these were the correct decisions. Only retrospectively did it become clear that the delay on the Mark V killed the project by postponing the decision to build until 1970 when, as already mentioned, political events and inflation intervened.

On the other hand I have no regrets about the bargain I made with the Council at that 1967 Christmastide. It was agreed that if I delayed the Mark V, then urgent priority would be given to a modification of the Mark I. This request had a far more practical and sinister basis than the desire to improve the performance. By that time the Mark I telescope had been in continuous use for ten years, and although a very high standard of maintenance had been achieved, it had been impossible to prevent some deterioration in the foundations on which the azimuth railway track was secured, and to the structure.

Over the years Husband had given me many warnings about the dangers. These warnings began after the winter of 1963 when some parts of the concrete work in the azimuth turntable showed signs of distress. It was about this time that Husband began his series of reminders that an instrument of this

[1] For the history of these negotiations about the 150-in. Anglo-Australian Optical Telescope see my paper in Quarterly Journal of the Royal Astronomical Society, 26, 393, 1985.

nature could not be expected to work forever merely with the daily maintenance which we were able to provide and that we should soon make provision for a large-scale overhaul. When the difficulties with the azimuth track arose Husband became anxious to make arrangements to reduce the load on this track. His warnings soon began to increase in urgency. He said that he must advise us to draw up a programme for improving the Mark I at an early date, since the telescope had been running for a very long time for what was 'virtually a prototype'. On 11 February 1967 Husband wrote again to say that he understood from our Chief Engineer that four more anchor bolts had broken where the cement packing had deteriorated under the rail baseplates, and he repeated his recommendation that the telescope should be modified to relieve the track and foundation loading. In fact, without any serious impact on the budget, we arranged during the course of 1967 for a part of the railway track to be relaid. Then, in September 1967, a far more serious situation was revealed during a routine inspection by one of our Shift Engineers. He noticed a thin line of rust on one of the massive steel cones which carry the whole load of the bowl to the trunnion bearings at the top of the towers (see Figures 19 and 20). The removal of this rust revealed a thin hair-line crack and soon more were found in the other cone.

Husband recommended that we should reduce the maximum speed in elevation to no more than 5 degrees per minute and avoid all sharp accelerations. After further inspection he wrote that there was little doubt that the cracks were due to fatigue caused by the alternating stresses set up by the elevation motion of the bowl. He warned me that, in a period of time which he could not estimate, the racks would eventually become entirely disconnected from the cones and would be held to the bowl structure only by the secondary connections. If we then still continued to use the telescope these secondary connections would suffer fatigue and fail, and the rack would fall out of mesh, probably causing tooth damage and perhaps jamming the bowl against any further rotation.

There is no doubt that I was thoroughly frightened by this letter, and this was the new and urgent operational background which encouraged me to take the instant decision in December 1967 that we should strike a bargain about proceeding with Ryle's telescope, and at the same time give priority to the Mark I modification at the expense of the delay which we had then thought to be two years in the construction of the Mark V.

No one seriously dissented from this proposal. The Research Council shed a budgeting dilemma and for the estimated cost of modification—£350,000—we would achieve a telescope that, according to Husband, would cost £2.5 million to build anew. By the summer of 1968 the Science Research Council made a public announcement of a grant of £400,000 for this work but any expectation I might have had that the modifications would proceed quickly and without hindrance was soon undermined by a sequence of events paralleling the problems of the original construction.

A series of re-designs

The formal application for this grant was made to the Science Research Council in February of 1968. The estimate was based on the 1964 proposals of Husband in which we had planned to construct a new and more accurate reflector of the same size and shape and held seven feet above the original reflector. Since the deflection of the single mast holding the equipment at the focus would be too great for the shorter wavelengths on which we intended to work, we planned to replace it by a tetrapod with the four legs spread 120 degrees diagonally and attached to the main structure above the points where the new double load-bearing wheels would take part of the load down to a new inner railway track. However, when this scheme was considered again early in 1968 we realised that the distribution of the legs of the tetrapod would scatter the incoming radiation and spoil the symmetry of the polar diagram of the telescope. In the light of researches on the polarization of the radio waves that had become significant in the four years since this plan was made, we no longer felt that this scheme was satisfactory. As soon as we presented this problem to Husband a whole series of difficulties arose. We could not find another more satisfactory arrangement of the tetrapod legs and soon decided to abandon the scheme in favour of retaining the single focal tower, but strengthened, so that the focal point would not deflect beyond acceptable limits as the elevation of the telescope changed. It was on this basis that Husband agreed that a figure of £350,000 for the modifications would be adequate and this led to the grant of £400,000 from SRC to cover the agent's fees and other subsidiary matters.

This plan, with the news of the award of the grant, was made public at a Press Conference convened at Jodrell on 9 July 1968. Alas, by that time more serious difficulties had emerged in the scheme for modifying the telescope. To celebrate the publication of this original book *The Story of Jodrell Bank* the Oxford University Press held a party in the University on 27 May. As he was leaving at the end of the evening Husband handed me a rough sheet on which he had made a sketch of the telescope bowl with the remark "We can give you a bigger telescope". I had frequently discussed with Husband the possibility of increasing the diameter of the telescope as part of this modification. The edge of the existing bowl was 45 ft. above the ground at zero elevation and in principle, at least, there seemed no reason why an increase in diameter should not be possible. However, Husband had always raised some objection from an engineering point of view, a prominent one being that this would upset the arrangement whereby the centre of gravity of the bowl structure was on the trunnion axis.

Now, with the prospect of going out to tender, Husband had begun to establish the details of the new load-bearing circular girders and had discovered that the weight of the steelwork in these girders could not be balanced by the proposed new reflector mounted only seven feet above the existing membrane. The solution he proposed, as shown in this sketch, was to increase the diameter of the new reflector to 265 ft. (that is 15 ft. greater than the original) and make the shape to be a long focus paraboloid with the focal point 82.15 ft. from the apex, instead of the existing 62.5 ft. As far as we were concerned this was an admirable arrangement. At the time of the Press conference in July we had arrived at what now seemed to me to be a typical telescope position—namely the Research Council and the Treasury had awarded us a grant and given their approval to a project that we were in process of re-designing with Husband. A week after the Press conference I received Husband's detailed report on the new scheme with the 265-ft. aperture reflector. From the engineering and scientific aspects everything seemed fine—we would get a 265-ft. aperture telescope instead of the present 250-ft. aperture, the surface would be far more accurate for use on short wavelengths and the worrying loadings on the trunnion bearings would be alleviated. There was, however, a major shock. Although tenders had not yet been sought the renewed estimate of the cost was at least £25,000 more than the amount of the grant so recently approved by the Research Council and the Treasury.

A few days later the Committee appointed by the Research Council to process the work met, with an official of the Council as their Chairman. My expectation that this would be a stormy meeting proved to be correct. Once more I had become a party to the all too familiar confrontation between the administrators from the London office and the men who had to modify and operate the telescopes. The Chairman demanded that we should again re-design the telescope to bring the cost of the modification within the scope of the grant already awarded. Husband said there was no possible 're-design' which would bring down the cost and that in any event a precise estimate could not be made until tenders were obtained. Eventually it was agreed that Husband would seek tenders to ascertain how much the steelwork *would* cost, but my hopes of getting the telescope modified during the spring and summer of 1969 were fast vanishing. There was a minor consolation. The tenders for the construction of the new load bearing inner railway track were within the estimates. The firm who sank the piles for the outer tracks (Chapter 11) were again employed and they began work at the end of September. At the end of the day's work we asked them to remove their rigs to that we could use the telescope at night. By November they had sunk 96 piles to the Keuper Marl, 70 ft. below ground level in this part of the site. The concrete top work followed and by early February of 1969 this new inner railway track was laid.

We were already referring to Phases 1 and 2 of the conversion. Phase 1 was now complete and Phase 2—the major structural work—was scheduled to take nine months from April to December 1969 during which we could have no use whatsoever of the telescope. This soon became a timetable without the slightest hope of achievement. In mid-December 1968 we had the devastating news of the tenders for the steelwork. The lowest was more than £100,000 greater than the estimate that had already caused the problem at the July meeting of the Committee. Altogether we now needed £504,000 (excluding fees) against the £350,000 of the original grant. I could have sought an additional grant from the Research Council but I did not want to do this. This was the time when the large Mark V telescope was merely 'postponed' and I feared that a request for more money for this conversion of the Mark I would undermine confidence in the estimates for the Mark V and would further jeopardise the chance of building that larger telescope.

It would be tedious to recount the succession of discussions that followed in the attempt to reduce the estimate to the level of the grant. By January 1969 we had reluctantly agreed to give up the idea of the enlarged reflector and return to the scheme with a new 250-ft. membrane. To achieve the balance, so that the centre of gravity remained on the trunnion axis, this new reflector, of poorer quality than that originally intended, would be mounted further above the original and there would be a heavy strake at the perimeter (that is the circular rim joining the old and new membranes around their circumference). If I could have foreseen the future and the ultimate decease of the plans for the Mark V and then the Mark VA telescope I would have had no hesitation in asking for more money to pursue the modification with the 265-ft. aperture bowl. However, no one could visualise at that time the subsequent rampant inflation in costs that was to lead to the cancellation of these projects and so we proceeded to seek tenders for this revised design.

Husband was understandably annoyed at these vacillations about the design of the modifications. Neither he nor I had much sympathy with the agents and administrators who were delaying the progress by such detailed scrutiny of every move we tried to make. Even more important was that we were now in mid-winter, two years after the decision to proceed with the changes to the telescope and, although the inner railway track was complete, absolutely nothing had been done to alleviate the strains and stresses on the telescope. Husband constantly warned us of the dangers of delay. In mid-January of 1969 he wrote to the agents who were acting on behalf of the Research Council "What prevents me from sleeping soundly on windy nights is the possibility of a collapse of the reflector structure or a failure of the elevation drive which would almost certainly completely destroy the telescope". For me, in the depths of those winter nights the telescope seemed to be transformed into a fragile instrument for which every breeze became a menacing hurricane.

By February 1969 my hopes that it would not be necessary to seek more money for the modifications were completely dashed. Prices were beginning to rise, the steel firms were busy and eventually only two of the many firms approached were even willing to submit tenders—and one of these, United Steel, the firm who built the original telescope, withdrew in favour of a major contract elsewhere. Eventually we were forced to accept the tender of the one remaining firm—Teesside Bridge—and by the summer of 1969 when all the costs had been assessed I had to apply to the Research Council for an additional grant of £145,000.

In the meantime our misfortunes increased. It was a severe winter, and the main concrete beam on which the telescope rotated suffered serious deterioration. The railway track had to be completely relayed and this added further to the financial burden. A new timetable became inevitable since, even if the money was available, Teesside Bridge could not begin work until eight months after the receipt of contracts. We set a new date of April 1970 for the beginning of this work, but in the meantime we were able to proceed with the procurement of the bogies for the inner railway track, the steelwork around the diametral girder, and the upthrust units on which the new heavy load-bearing circular girder wheels would run. Phase 2 became Phases 2 and 3, and the money for Phase 3 was not yet available.

A fortunate compensation

There was one fortunate compensation for these delays. If my original hopes had materialised the major steelwork changes would have occurred in the spring and summer of 1969 and we would have been immobilized during Man's first attempt to land on the moon. Apollo 11 was launched from Cape Kennedy on 16 July, and although the Mark I telescope was not directly involved in the tracking of this vehicle, we did not escape the drama of the occasion. The Soviets had launched their Luna 15 vehicle three days earlier and the Mark I followed the course of this probe in detail as it was placed in orbit around the moon. The Americans were concerned and sought assurance from the Soviet authorities that no changes would be made in the orbit of Luna 15 or that it would in any way compromise or interfere with the manned Apollo 11. Since our records showed that the orbit of this lunik was quite different from any previous one, we believed that the Soviets would attempt to make a soft landing of the vehicle and recover lunar samples to Earth automatically without hazarding the lives of human beings. Throughout the landing of Apollo 11, and most of the twenty-one hours that Armstrong and Aldrin spent on the lunar surface, the lunik continued in orbit around the moon. Then, shortly before Apollo 11 was due to lift-off on 21 July, the rockets of the lunik were fired and it crashed to the moon in the general vicinity of Apollo 11. Our belief that this was an attempt at a soft landing which failed was strengthened immediately when we received a telegram from the President of the Soviet Academy asking for our data on the last minutes of the flight of this vehicle.[1] Confirmation came over a year later, in September 1970, when the Soviets succeeded with Luna 16 in making an automatic recovery of lunar soil.

The winter of 1969–70

These excitements were over when on 3 September 1969 we handed over the telescope to the contractors for the re-laying of the railway track. By mid-November this work was complete, and four new bogies for the inner track and the steelwork around the diametral girder were in place. Once more we could have the telescope for our research programmes. It was an odd situation. We had spent £200,000 but from the practical point of view we had accomplished nothing of the load relief for which the grant application was made in February 1968. Now, throughout the winter of 1969-70 we were able to use the telescope in its original form with the addition of the four bogies being carried around on the inner track, but not yet bearing any of the load for which they were destined.

This was a winter of gloom and anxiety. Gloom, because Teesside, having been awarded the steelwork contract, announced that their work would take eleven months instead of nine, and if they began in April of 1970 it would be the end of February 1971 before the work could be completed. Anxiety, because in November our shift engineers discovered further cracks in sections of the main framework carrying the bowl to the trunnions. So we limped through our winter's research programme, never using the telescope in winds of more than 25 mph and with maximum speed of movement in elevation reduced by one-half.

The final phase

Throughout the winter and the early spring of 1970 the proposed starting date of April 1970 constantly receded and nothing I could do improved the position. Every kind of difficulty was raised, from delays in delivery dates of the structural steel to lack of agreement on the procedure for lifting the new membrane sections from the ground over the edge of the existing bowl. The intended starting date of 6 April passed, and not until the 14 August 1970 did we drive the Mark I to the zenith for the last time so that the Teesside Bridge engineers could begin their work. The Mark I had worked almost continuously for thirteen years and life without it attained a strange unreality. At least we could look forward to the use of the modified telescope—the Mark IA—in June 1971 and began to occupy ourselves with whatever research was possible

[1] The Academy wanted these data so urgently that they were not satisfied by my response that we would post the data to them. Instead they sent a special emissary to Ringway (Manchester) airport the next day to whom we handed copies of our records.

with the remaining Mark II and III telescopes. The overall plan, with the bowl locked in the zenith, was to build the load-bearing wheels first so that the membrane erection could commence in February 1971. Soon there were a series of major problems with sections of the fabricated wheel girders. In October 1970 a flange on one of the girders being fabricated at Teesside Bridge had cracked. The reason for this was unknown until we had the report of the metallurgical examination in January 1971. To our dismay, the conclusion was that the British Steel Corporation had supplied a batch of faulty steel on which the correct normalising treatment had not been carried out. The elimination of this faulty steel from that being used to build the wheel girders was a time-consuming business. Shortly after we had this report, one of the good wheel girder sections was accidentally cut through with an oxy-acetylene torch at the Teesside Bridge works, so that by mid-February the erection of the girders on the telescope was thirteen weeks behind schedule.

Even when the fabricated sections arrived at Jodrell the work of erection seemed to me to be proceeding without any sense of urgency, and I complained continuously to Husband and to the agents. Towards the end of April 1970 Husband told me he could not speed up the work. A major problem, he said, was that we were dealing with a nationalized industry which did not always respond as quickly as one would wish. In May 1971, only a few weeks from the original date for the completion of the whole conversion, the wheel girder structure was completed. The upthrust bogie units, on which the wheels rested, were jacked up to give a load relief of 170 tons on the trunnions and at last on 20 May the erection of the supporting steelwork under the old bowl, to hold the new membrane, commenced.

The problem of lifting the new membrane sections more than 200 ft. from the ground over the edge of the original rim of the telescope bowl was solved by building a single tower crane of sufficient height and lift. This moved on a short railway track, crossing the main circular azimuth track of the telescope. With this crane outside of the telescope azimuth track, the telescope could be rotated and then the crane would move in and lift the membrane sections over the rim and place them in their final position on the new surface. The difficulty of the task of fixing this new membrane in position had been seriously underestimated. In September, when we asked the Manager of the work force why, with the progamme so far behind target, the working day had been reduced by one hour, we were advised that the erectors working underneath the new membrane panels had to cling to the steelwork all day and found this very tiring. Indeed, like the original construction of the Mark I, it was a formidable task and the last panels were not placed in position until the end of October 1971. So, not in 1969, nor in 1970, but in mid-November 1971 we were at last able to claim some use of the telescope. In 1957 the Mark I had been pressed into use to locate by radar the carrier rocket of the Sputnik. Now, 14 years later, the Mark IA could track two Soviet probes to the planet Mars[1], but, as in 1957, month after month passed before we could clear the site of contractors and painters.

The final cost?—£664,793.

The primary incentive for the modification had been the anxieties about the safety of parts of the structure of the Mark I. The conversion not only overcame these problems but produced a far better instrument for astronomical research. When the membrane panels had been adjusted, the paraboloidal surface worked on a wavelength of 6 cm with an efficiency of 30 per cent. and on the important hydrogen spectral line wavelength of 21 cm, the efficiency was close to the maximum possible as we had hoped.

From the critical decision, taken in the office of the Chairman of the SRC in December 1967, to the signature on the handover documents, the conversion of the Mark I to the Mark IA had taken 6.5 years—as long as the construction time for the Mark I and costing more than the original instrument. During this long period, and especially when we were without use of the telescope in 1970 and 1971, I had repeatedly complained to Husband about the delays. To my surprise he replied that it was the most difficult engineering job he had ever tackled. As the costs of the work soared and as the likelihood of attaining the Mark V project became more and more improbable, Rainford[2] remarked that the decision to do the conversion was the wisest one we had ever made.

[1] Mars 2 reached the planet on 27 November and Mars 3 on 2 December 1971.

[2] R. A. Rainford, who figures so prominently in the original story of the Mark I, retired from the office of Bursar of the University in 1971 but retained a close connection with the Mark IA, V & VA affairs and was appointed Chairman of the Project Committee overseeing the construction of the MERLIN system (see Chapter 40).

40

MERLIN—and the future

In 1972, when once more we had full use of the Mark IA telescope (the Mark V 400 ft. project having been abandoned), we were in the throes of designing the 375-ft. Mark VA. The rapidly rising costs and the negative attitude of the administrative and financial bodies to large projects soon combined to convince me that we were unlikely ever to build this telescope. We had wasted enough time and in the last months of 1973 I had many discussions with the senior staff at Jodrell to see if we could prepare the background for an alternative scheme that would keep Jodrell in the forefront of astronomical research. The Mark IA was working splendidly, heavily engaged with the researches on radio galaxies, quasars, pulsars and with the high efficiency at the shorter wavelengths, now proving to be a powerful instrument for the investigation of the neutral hydrogen and hydroxyl clouds in the Milky Way.

As related in Chapter 38, we had pioneered the development of long baseline interferometers for the angular diameter measurements of the radio galaxies and quasars. During the course of this work it had become apparent that the radio sources rarely had a symmetrical structure and the significance of mapping the sources was underlined by the work in Cambridge, where Ryle and his colleagues had developed the technique of aperture synthesis. In this technique the resolution equivalent to a very large single telescope could be synthesised.

The idea was straightforward. If one had a steerable paraboloidal telescope with an aperture of, say, one mile, then one could direct the beam over all angles of the radio source and so obtain a detailed map of the structure. Although one cannot build a radio telescope with a reflecting bowl one mile in diameter, it is possible by using two smaller telescopes to 'fill in' this aperture of one mile, to carry out large numbers of observations of the source over many days, with the telescopes in all possible positions. Provided the radio source itself does not vary over the period of observation and provided that the radio information from the two telescopes can be stored so that the phase of the signals is preserved, then by using the appropriate computer analysis it is possible to combine these many individual sets of observations and synthesize the radio image that would have been obtained by using a single one-mile aperture paraboloid.

It was the great success of this development in Cambridge that led me in 1967 to agree to delay the Mark V telescope so that Ryle could build the Cambridge 5-km aperture synthesis system. A similar system was under construction in Holland and in the 1970 decade both systems produced many detailed radio maps revealing the complex structure of the emission from distant galaxies and quasars. However, the resolving power of these systems was in the seconds of arc region. Although this produced most interesting maps of the radio galaxies and quasars, the resolution was too poor to give any details of the central cores where the immense energies were generated. Already in 1966, as I have related in Chapter 38, our own long baseline interferometer had shown that the cores of some of these radio sources had an angular extent of less than a few hundredths of a second of arc.

The VLBI (Very Long Baseline Interferometer) development

In September 1966 I sought a grant from the Research Council so that we could develop an interferometer system with the aerials ten times further apart than the 127 km of the Jodrell-Defford baseline. In this system separate tape recordings would be made at the individual sites. Using magnetic tape recorders, atomic clocks, and computer techniques then available, we planned to bring the two tapes together after the real time recordings had been made and produce the interference pattern from which we could estimate the angular diameters as though the two aerials had been connected by radio links. By extending the baselines in this way we hoped to find out how much less than a few hundredths of a second of arc in angular extent some of these radio sources were. We had heard that Canadian radio astronomers were developing a similar system and had already had preliminary contacts with them about establishing a transatlantic baseline of several thousand kilometres. The Canadians (and then the Americans) succeeded in testing their systems long before we did. In the spring of 1967 the Canadians established baselines of over 3000 km and found the surprising result that some of the quasars were still unresolved. In the summer of 1967 a group of American radio astronomers found that out of twenty-eight sources studied, eleven had an angular extent of less than 0.01 seconds of arc, and a new limit of 0.005 seconds of arc was established for some of the quasars. By 1968 this remote linkage of telescopes in America and the Soviet Union—a baseline of over 8000 km established that there were quasars whose angular extent was less than 0.0006 seconds of arc. It was 1969 before our own system operated successfully and then with the Mark I we were able to establish the first transatlantic interferometer with telescopes at the Algonquin and Penticton telescopes in Canada. Even with these multi-million wavelength interferometer baselines there were still many quasars and radio galaxies that could not be resolved. Thus, the straightforward two-telescope VLBI systems had shown that structural detail must exist in those quasars with angular extent many thousands of times less than that shown with the resolution of the Cambridge and Dutch aperture synthesis instruments.

The MERLIN system (Multi-Element-Radio-Linked-Interferometer)

That was the background to a suggestion made to us by Henry Palmer late in 1973 when the Mark V-VA telescope projects seemed on the point of rejection. In principle he proposed that we should add more telescopes to the existing Jodrell-Defford interferometer so sited that when the signals from all the telescopes were combined we would have a radio source map showing detail of less than a second of arc. When the Mark VA proposal finally collapsed in 1974 we were ready to seek a grant from the SRC for this project. There were many technical difficulties and the concept was opposed by some of our radio astronomical colleagues elsewhere on the grounds that we would not be able to make a workable system or that the fine structure which we were seeking in the radio objects either did not exist or was not of sufficient interest to make the research worthwhile. The grant I was seeking—less than £4 million for this system—was only a third of the sum earmarked in the Research Council's budget for the Mark VA and it was intensely annoying and frustrating that this misguided opposition from people for whom we had made such sacrifices succeeded in delaying the work to such an extent that it was the end of May 1975 before we had authority to proceed—and then only with a scaled-down version on the grounds that we must first prove that our ideas were correct and workable.

I have published a detailed account[1] of the development of the two phases of this system into which we were forced by this unscientific opposition. We wanted additional telescopes of about 80 ft. aperture and after my initial contacts with Husband it soon became clear that by far the most economical means of proceeding would be to purchase telescopes from an American source. Our friendly contacts with the American National Radio Astronomy Observatory soon led to an agreement whereby we purhcased telescopes from E-Systems, Inc., of Dallas. This firm was in process of manufacturing 28 telescopes of 82 ft. diameter for the American VLA (Very Large Array) being built in New Mexico. Eventually E-Systems supplied and installed three of these telescopes in our network with such efficiency and speed that they were available for use long before our own complex electronic developments were ready.

The first phase used one of these new E-Systems telescopes situated 68 km south-west of Jodrell. It was computer controlled from Jodrell and the radio signals were transmitted over radio links to a computer and analytical system

[1] The Jodrell Bank Telescopes, O.U.P. 1985. Chapter 17 et seq.

in the Control Building at Jodrell. With this telescope, the Mark IA, Mark III and the Defford telescope all tracking the same quasar the signals from all the pairs of telescopes were correlated to produce an intensity contour map of the quasar. By the autumn of 1977 we had produced incontrovertible evidence that the various objections raised to the original proposal were invalid and in June 1978 we were given the grant to complete the second phase of the system. This involved the construction of two more E-Systems telescopes—11 km north-west of Jodrell and 18 km south-west. As 1980 was drawing to a close the complete system came into operation. Six telescopes operating simultaneously on the same area of sky began to produce maps of a quality far beyond that originally envisaged when the system was proposed six years earlier.

As I write, towards the end of 1986, the MERLIN system has mapped nearly 700 radio galaxies and quasars. It is typical of the contemporary computerised scientific equipments that the actual observation of the heavens by the telescopes is a small part of the process needed to obtain the final result. MERLIN is no exception and so far the final contour maps are available for only half of those for which the data are on the tapes.

An example of the radio maps obtained is shown in Figure 39. This is of the quasar 3C 273, referred to in Chapter 38. It was the determination of the precise position of this radio source by Hazard, using the lunar occultation technique in 1962-63, that first led to the identification of the quasars as distant extragalactic objects. On the photographic plate 3C 273 appears as a star-like object accompanied by a faint wisp. The MERLIN radio map shown here has a resolution of 1 arc-second and reveals the complex radio structure of the jet emerging from the unresolved core. However, subsequent observations on an international basis have enabled the core to be studied with the maximum terrestrial baselines possible. At resolutions more than a thousand times superior to that of the map in Figure 39 there remains an unresolved core with a jet emerging—but not colinear with the outer jet of this Figure. These milliarc second limits to the extent of the inner core mean that, in a quasar like 3C 273, energy equivalent to the total annihilation of the mass of a million suns is being generated within a volume of space only about 2 parsecs in extent. In our own neighbourhood of the Universe we have no energy sources bearing comparison with this phenomenon. The Sun, like the stars, generates energy by the thermonuclear process involving the conversion of hydrogen to helium. In these processes only a percent or so of the mass is converted into energy. For comparison with the core of a quasar we have to think in terms of the total annihilation of the mass of some millions of suns within a volume of the universe no greater than that encompassing the Sun and the nearest stars.

In addition to the map-making of the radio galaxies and quasars MERLIN has been used in a variety of researches. Spectral line maps have been made of hydroxyl and water vapour sources in the galaxy, and the system has been used in astrometric measurements, for pulsar searches and for measurement of the motion of the pulsars.

By the time we drove the Mark I telescope to the zenith in August 1970 for the conversion to the Mark IA it had been used for 68,538 hours for our research work. Towards the end of 1971 the converted instrument, now known as the Mark IA, came into use. Now, fifteen years later, a further 80,000 hours of research have been carried out with the telescope. There were those who thought that the advent of MERLIN would lessen the interest in the Mark IA because they believed that the smaller but more accurate Mark II would be the more suitable home-based telescope to be incorporated in the network. On the contrary, the demand for the telescope has never been so great as it is today. Nearly half of the operational time of the Mark IA has been either in the localised Jodrell MERLIN network, or as a component of the European-transatlantic VLBI network. The other half of its use has been on the hydroxyl and neutral hydrogen spectral line researches, on pulsars and many other miscellaneous researches—including the occasional involvement in the tracking of deep space probes.

The near destruction of the telescope

The survival of the telescope as a front line instrument for research over a time span of thirty years is a phenomenon of our age. Scientifically, a twenty-four hour schedule has never been able to satisfy those who want to use it. As an engineering structure there have been few problems. Nevertheless, on the 2nd January 1976 I was given a sharp reminder that, like a ship at sea, the telescope was in constant danger from the elemental forces of nature. It was my habit to leave Jodrell at about 6:30 p.m. and drive the few miles to my home in a nearby village. Always I would make a final visit to the control room to have a few words with the Controller, who would be on duty until his shift ended at midnight. I did so on 2 January 1976. The telescope was already locked in the zenith because a high wind had been blowing all day. I was not unduly worried since a check over the direct telephone line with the meteorological officer at Ringway airport indicated that we were not likely to suffer gusts of more than 60 mph. With the azimuth of the telescope adjusted so that the towers were in line with the wind direction this was not a matter of great concern and I proceeded on my homeward journey. The few miles of lanes through which I drove were scattered with twigs and small branches from the trees and I was thankful to park the car and reach the comfort of the fireside. About an hour later the noise of the wind began to worry me. A great tree crashed to the ground near the house and as I reached for the telephone which would connect me with the Jodrell control room the Controller phoned to urge me to return to Jodrell. The shift engineer had reported that the huge wheel girders had been shifted more than seven inches across the upthrust units by the force of the wind. Another inch and the entire one and a half thousand ton weight of the bowl structure would fall on the trunnion bearings and the whole elevation structure of the telescope would probably collapse. The Chief Engineer was on leave and the Controller had already telephoned his deputy but his route to Jodrell was completely blocked by fallen trees.

With the greatest difficulty I fought against the wind to reach my car, only thirty yards from the house. The village seemed deserted and not a single vehicle or person was to be seen on my route to Jodrell. I reached the Control Room to find that the mean wind speed was over 70 mph and still rising and the pen of the barometer was falling as though over a steep cliff. The shift engineer and the Controller were in a state of alarm. By this time only half an inch of the wheel girder on the bogies of the upthrust units was relieving the trunnion bearings of an insupportable load. There was only one possible solution—to rotate the telescope through 180 degrees so that the great force of the hurricane would shift the wheel girders back on to the upthrust units. In doing so the telescope would become athwart the wind and thus move through a condition for which the wind rakers and outer bogies had never been designed or subjected. With the wind now over 90 mph and reaching hurricane force 12 we switched to our diesel generators in case the main supply should fail at a critical stage and set the telescope in slow azimuth motion. After some 45 degrees of rotation I saw with immense relief the sudden jump of the wheel girders as they slid over the upthrust units to their normal position.

By the time we had completed this move the worst of the storm was over and the telescope was safe. By daybreak there was still a gale force eight to nine but we were able to inspect the vital parts of the telescope structure. There was no obvious damage. Clearly the telescope had been built to survive. I discovered later that Husband was far away from the centre of this hurricane and had no fear on that night for its safety. When he eventually analysed the problem he concluded that the great force of the wind on the upturned bowl had actually tilted the towers from the vertical and thereby shifted the whole bowl transversely so that the load bearing wheel girders had nearly been removed from the upthrust units fixed on the diametral girder. In the Mark I the 'bicycle wheel' was merely a stabilising device. It carried no load and such a shift would have been of no consequence. Now in the Mark IA modification a significant part of the 1500 ton load of the bowl structure was taken to the inner railway track through the wheel girders. Fortunately the solution was relatively simple. Four steel girders were inserted as diagonal braces of the two towers to the diametral girder. They can be seen in the photograph of the Mark IA telescope (Figure 40).

The years ahead

University Professors in England retire at the age of 67, so at the end of September 1980 I vacated the Chair of Radio Astronomy. For a further year I remained the director of Jodrell Bank to enable my successor to complete his duties elsewhere. It was my great good fortune that the person who succeeded me as director at the end of September 1981 was an old friend and colleague. Graham Smith was one of the pioneering radio astronomers in Ryle's group at Cambridge whom we appointed to a Manchester chair in 1964 to work with us at Jodrell. In 1974 when all hopes of the Mark V/VA telescope had vanished he transferred to the Royal Greenwich Observatory at Herstmonceux as the director, and now, in 1981, he returned to take over the directorship of Jodrell and the responsibility for the telescopes from me.[1] Thus my account of the immediate future plans at Jodrell are of the plans made by him and his staff. It is a thrilling prospect likely to be tarnished only by a lack of vision of those from whom the finances have to be sought.

The first step in these future plans is to add another telescope into the MERLIN network. A series of radio links have already been established to the site of the radio astronomy observatory at Cambridge and in December 1986 an existing 60-ft. telescope was added to the network. This will provide the longest baseline of 233 km between the telescopes of the network and still further improve the resolution of the maps. The additional six pairs of interferometric baselines will greatly improve the detail of the contour maps and will also fill in a critical parameter in the European VLBI network. Unfortunately the telescope at Cambridge is smaller than the other remote telescopes in the MERLIN network. Therefore it is planned to install a 100-ft. aperture telescope on that site so that the network will operate at its full sensitivity.

When the Mark IA telescope is facing towards the east I look straight into its huge bowl from my office. Like the whole telescope it has a regular painting schedule but this steel membrane is unprotected from the heat, cold, rain or snow, and I see patches of discoloration. Significant sections of it are rusting. The unfortunate harvest of the necessity to build it of poorer quality material than intended during the Mark I to IA conversion is now being reaped. It may well last for a few more years but must soon be replaced. When this is done it is not planned merely to replace the rusting sheets of steel. Plans exist for building in a sophisticated arrangement of jacks so that the surface can be maintained as a very accurate paraboloid by remote computer control. If this plan materialises the telescope would then be capable of use at very short wavelengths in regions of the spectrum where our knowledge of the universe remains fragmentary.

In Chapter 16 and the succeeding chapters I described the problems with the driving system of the telescope. The final solution (Chapter 25) was the

[1] Shortly after his arrival at Jodrell he was appointed Astronomer Royal (the title being no longer tied to the director of the Royal Observatory). In 1985 he was knighted and became Sir Francis Graham Smith.

installation of the Ward Leonard system by Messrs Brush. Over the years, apart from the need to replace a bearing, this system has worked with great reliability. However, thirty years after the installation this massive arrangement of rotating machinery was regarded as old fashioned and very wasteful of power. In 1985 the azimuth motorised system in the Ward Leonard room at the centre of the diametral girder was replaced by a modern solid state system. This has worked satisfactorily and when the elevation drive system is similarly modified the familiar roar of the machinery will disappear and to the onlooker the mystery, not only of the smooth, but of the silent movement of these thousands of tons of steel will deepen.

When I first walked through the gates of Jodrell Bank more than forty years ago, astronomers using the world's great optical telescopes had penetrated far into space and time. Much was known about the stars and galaxies but there was an underlying mystery about the universe. In these forty years the radio telescopes, the earth satellites and the space probes have enabled us to penetrate further into time and space and there has been a vast accumulation of new knowledge. But the mystery remains. Nearly every problem that has been solved has revealed an even deeper lack of understanding of the nature of the cosmos and of Man's relation to the elemental forces that ultimately led to the emergence of life on Earth. When the Jodrell telescope came to life in 1957 I was full of optimism and expectation that, at last, we had the device that would soon solve the remaining major problems about the nature and evolution of the universe. Indeed, it may be true that the major issues of 1957 have been settled but the ultimate mystery remains unsolved. Can a solution be discovered by the use of telescopes and space probes? I have reached the stage where I do not believe that there is a solution through these physical avenues of exploration. The next forty years at Jodrell will be as exciting and thrilling as those I have lived through myself since that winter's day in 1945 when I arrived at these gates. No doubt, by then, the problems of today will be solved, but the task of comprehension has been and will remain that of every generation of mankind.

Index

ABOUT THE AUTHOR

Sir Bernard Lovell, O.B.E., LL.D., D.Sc., and Fellow of the Royal Society of London, is the former professor of radio astronomy at the University of Manchester and director of the Experimental Station at the Nuffield Radio Astronomy Laboratories, Jodrell Bank, Macclesfield, Cheshire, England. He is the distinguished author of several books and many articles dealing with space exploration, the nature of the universe, the origins of life and man's relationship to the cosmos. He also lectures widely on the implications of scientific growth and its consequences for man and the universe, showing with compelling clarity and originality the convergence of man with nature.

ABOUT THE FOUNDER OF THIS SERIES

Ruth Nanda Anshen, Ph.D., Fellow of the Royal Society of Arts of London, founded, plans, and edits several distinguished series, including World Perspectives, Religious Perspectives, Credo Perspectives, Perspectives in Humanism, the Science of Culture Series, the Tree of Life Series, and Convergence. She also writes and lectures on the relationship of knowledge to the nature and meaning of man and to his understanding of and place in the universe. Dr. Anshen's book, *The Reality of the Devil: Evil in Man*, a study in the phenomenology of evil, demonstrates the interrelationship between good and evil. She is also the author of *Anatomy of Evil* and of *Biography of an Idea*. She has lectured in universities throughout the civilized world on the unity of mind and matter and on the relationship of facts to values. Dr. Anshen is a member of the American Philosophical Association, the History of Science Society, the International Philosophical Society and the Metaphysical Society of America.